# Oxford Mathematics 6
Primary Years Programme

# Contents

## NUMBER, PATTERN AND FUNCTION

### Unit 1 Number and place value
1. Place value — 2
2. Square numbers and triangular numbers — 6
3. Prime and composite numbers — 10
4. Mental strategies for addition and subtraction — 14
5. Written strategies for addition — 18
6. Written strategies for subtraction — 22
7. Mental strategies for multiplication and division — 26
8. Written strategies for multiplication — 30
9. Written strategies for division — 34
10. Integers — 38
11. Exponents and square roots — 42

### Unit 2 Fractions and decimals
1. Fractions — 46
2. Adding and subtracting fractions — 50
3. Decimal fractions — 54
4. Addition and subtraction of decimals — 58
5. Multiplication and division of decimals — 62
6. Decimals and powers of 10 — 66
7. Percentage, fractions and decimals — 70

### Unit 3 Ratios
1. Ratios — 74

### Unit 4 Patterns and algebra
1. Geometric and number patterns — 78
2. Order of operations — 82

## MEASUREMENT, SHAPE AND SPACE

### Unit 5 Using units of measurement
1. Length — 86
2. Area — 90
3. Volume and capacity — 94
4. Mass — 98
5. Timetables and timelines — 102

### Unit 6 Shape
1. 2D shapes — 106
2. 3D shapes — 110

### Unit 7 Geometric reasoning
1. Angles — 114

### Unit 8 Location and transformation
1. Transformations — 118
2. The Cartesian coordinate system — 122

## DATA HANDLING

### Unit 9 Data representation and interpretation
1. Collecting, representing and interpreting data — 126
2. Data in the media — 130
3. Range, mode, median and mean — 134

### Unit 10 Chance
1. Describing probabilities — 138
2. Conducting chance experiments and analysing outcomes — 142

Glossary — 146
Answers — 156

OXFORD UNIVERSITY PRESS
AUSTRALIA & NEW ZEALAND

# UNIT 1: TOPIC 1
# Place value

**Working with very large numbers**

Large numbers have a gap between each set of three digits.

837452691 is easier to read if we write 837 452 691. It also makes it easier to say the number:

eight hundred and thirty-seven million, four hundred and fifty-two thousand, six hundred and ninety-one

## Guided practice

**1** Look at this number: 5 367 918

Show the value of each digit on the place-value grid.

| Millions | Hundred thousands | Ten thousands | Thousands | Hundreds | Tens | Ones | Write the number using gaps if necessary |
|---|---|---|---|---|---|---|---|
| 5 | 0 | 0 | 0 | 0 | 0 | 0 | 5 000 000 |
|   |   |   |   |   |   |   |   |
|   |   |   |   |   |   |   |   |
|   |   |   |   |   |   |   |   |
|   |   |   |   |   |   |   |   |
|   |   |   |   |   |   |   |   |

**2** If we write nine hundred and five thousand, four hundred and seventy-six in digits, we use a zero to show there are no tens of thousands:

9**0**5 476

*Remember to use a zero as a space-filler.*

Write as digits:

a  fifty-one thousand, six hundred and four  _____

b  two hundred thousand and twenty-six  _____

c  twelve thousand and ten  _____

## Independent practice

**1** What is the value of the red digit?

a  463 290  _____  b  6 329 477  _____

c  2 406 219  _____  d  51 385 067  _____

e  80 487 003  _____  f  351 000 819  _____

**2** Write the numbers from question 1 in words.

a  _____

b  _____

c  _____

d  _____

e  _____

f  _____

**3** Write these numbers as digits.

a  eighty million, four hundred and eighty-seven thousand

_____

b  ten million, three hundred and sixty-two thousand and fifty-nine

_____

c  one hundred and fourteen million, seven hundred and sixty thousand, two hundred and nine

_____

d  one billion, four hundred million, five hundred and ninety-three thousand and one

_____

OXFORD UNIVERSITY PRESS

**4** Expand these numbers. The first one has been done for you.

*Remember to use spaces between the digits where necessary.*

a 374 596:    300 000 + 70 000 + 4000 + 500 + 90 + 6

b 214 867:    200 000 + _____

c 2 567 321:  _____

d 5 673 207:  _____

e 57 319 240: _____

f 407 508 004: _____

---

**5** Look at these digit cards.    7   3   4   5   9   1   2

a What is the **largest** number that can be made using all the cards?

b What is the **smallest** number that can be made if the digit "5" is in the millions place?

c What is the **largest** number that can be made if the "7" is seven ones?

d What is the **smallest** number that can be made if the "1" is in the tens of thousands place?

---

**6** Write the number shown on each spike abacus as digits and in words.

a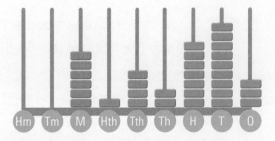

digits: _____

words: _____

_____

b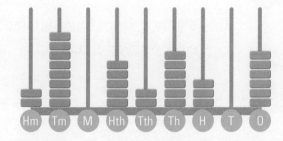

digits: _____

words: _____

_____

# Extended practice

**1** To change the calculator screen to show the second number, I would press:

a  `24 550`  _____ = `24 650`

b  `37 154`  _____ = `77 154`

c  `739 255` _____ = `719 255`

d  `999 999` _____ = `1 000 000`

---

**2** Sometimes large numbers are abbreviated. $1K means $1000. $1.3M can be used for $1 300 000. Write the new price of these houses using digits **in full**.

a  $345K reduced by $5000        _____

b  $725K reduced by $20 000      _____

c  $875K reduced by $50K         _____

d  $1.5M reduced by $250K        _____

---

**3** Imagine you have to choose just **one** digit in each of these numbers. Write:
- the digit you would choose
- the value of the digit
- the reason for your choice.

a  A share of $574 612. _____

_____

b  Writing out your times tables 574 612 times. _____

_____

c  Eating 574 612 of your favourite snack food in 10 minutes. _____

_____

# UNIT 1: TOPIC 2
## Square numbers and triangular numbers

**Numbers can be arranged in patterns**

"4 is a square number."

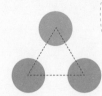
"3 is a triangular number."

## Guided practice

**1** These are the first six square numbers. Fill in the gaps.

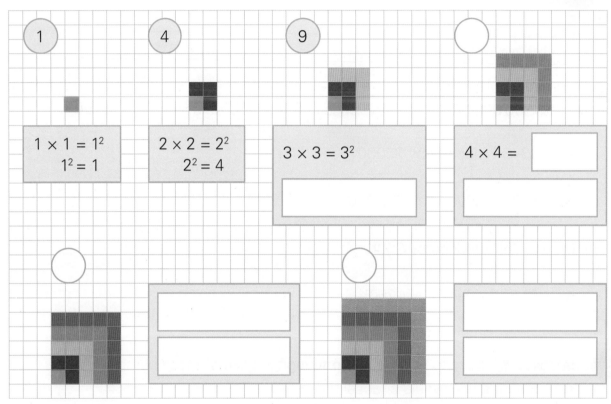

1         4         9         ◯

$1 \times 1 = 1^2$
$1^2 = 1$

$2 \times 2 = 2^2$
$2^2 = 4$

$3 \times 3 = 3^2$

$4 \times 4 =$ ☐

◯                    ◯

**2** These are the first four triangular numbers. Fill in the gaps.

1         3         ◯         ◯

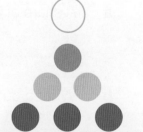

| 1 | $1 + 2 = 3$ | $1 + 2 + 3 =$ ☐ | ☐ |

6

# Independent practice

**1** Complete the grid to show the first ten square numbers. Write the information as you did on page 10.

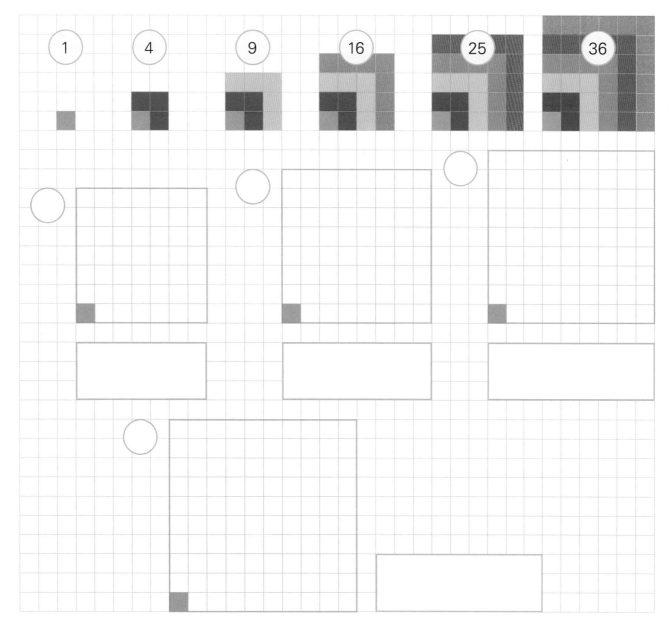

**2** a What is the next number in the square number pattern? ☐

b How does the digit in the ones column change in the square number pattern?

_____

c Circle one answer. The 100th square number is:

100     1000     10 000     100 000

**3** Complete the pattern and information to show the first 10 triangular numbers.

① 1    ③ 3    ⑥ 6    ⑩ 10    ○

| 1 | 1 + 2 = 3 | 1 + 2 + 3 = 6 | 1 + 2 + 3 + 4 = 10 | 1 + 2 + ___ |

○    ○    ○

[   ]    [   ]    [   ]

○    ○

[   ]    [   ]

**4** a  What is the 11th triangular number? [   ]

b  Apart from 1, which triangular number is also a square number? [   ]

c  How does the triangular number pattern grow? (Hint: Think about odd and even numbers.) _____

# Extended practice

**1** Continue this table.

| Square number | Multiplication fact | Addition fact |
|---|---|---|
| $1^2 = 1$ | $1 \times 1 = 1$ | 1 |
| $2^2 = 4$ | $2 \times 2 = 4$ | $1 + 3 = 4$ |
| $3^2 = 9$ | $3 \times 3 = 9$ | $1 + 3 + 5 = 9$ |
| $4^2 =$ | | |
| $5^2 =$ | | |
| $6^2 =$ | | |
| $7^2 =$ | | |
| $8^2 =$ | | |
| $9^2 =$ | | |
| $10^2 =$ | | |

**2**

**a** What do you notice about the way the addition facts grow in question 1? _____

**b** Write the facts for the 11th square number. _____

**c** How many would you add to the 11th square number to find the 12th square number? _____

**3** This pattern shows the first few pentagonal numbers.

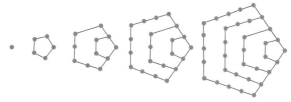

**a** One of the numbers in this list is **not** a pentagonal number. Which number is it?

5, 12, 15, 22, 35

**b** Write the first 5 pentagonal numbers. _____

**c** Write an explanation that would help a younger student to understand the connection between each pentagonal number and the one that follows it. _____

**d** On a separate piece of paper, draw a diagram of the 6th pentagonal number.

# UNIT 1: TOPIC 3
# Prime and composite numbers

**How do we recognise a prime number?**

We say a number is *prime* if it has just two factors: 1 and itself. The number 2 is the smallest prime number because it can only be divided by 1 and 2. Numbers that have more than two factors are called *composite* numbers.

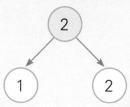
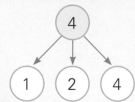

## Guided practice

*1 only has one factor, so it is neither a prime number nor a composite number.*

**1** Complete this chart.

| Number | Factors (numbers it can be divided by) | How many factors? | Prime or composite? Prime | Composite |
|---|---|---|---|---|
| 1 | 1 | 1 | neither | |
| 2 | 1 and 2 | 2 | ✓ | |
| 3 | | | | |
| 4 | | | | |
| 5 | | | | |
| 6 | | | | |
| 7 | | | | |
| 8 | | | | |
| 9 | | | | |
| 10 | | | | |
| 11 | | | | |
| 12 | | | | |
| 13 | | | | |
| 14 | | | | |
| 15 | | | | |
| 16 | | | | |
| 17 | | | | |
| 18 | | | | |
| 19 | | | | |
| 20 | | | | |

**2 a** List the prime numbers between 2 and 20. _____

**b** Comment on the number of even prime numbers. _____

## Independent practice

**1** Follow these instructions to complete the grid. The grid has been started for you.

| 1☆ | 2○ | 3 | 4 | 5 | 6 | 7 | 8 | 9 | 10 |
|----|----|----|----|----|----|----|----|----|----|
| 11 | 12 | 13 | 14 | 15 | 16 | 17 | 18 | 19 | 20 |
| 21 | 22 | 23 | 24 | 25 | 26 | 27 | 28 | 29 | 30 |
| 31 | 32 | 33 | 34 | 35 | 36 | 37 | 38 | 39 | 40 |
| 41 | 42 | 43 | 44 | 45 | 46 | 47 | 48 | 49 | 50 |
| 51 | 52 | 53 | 54 | 55 | 56 | 57 | 58 | 59 | 60 |
| 61 | 62 | 63 | 64 | 65 | 66 | 67 | 68 | 69 | 70 |
| 71 | 72 | 73 | 74 | 75 | 76 | 77 | 78 | 79 | 80 |
| 81 | 82 | 83 | 84 | 85 | 86 | 87 | 88 | 89 | 90 |
| 91 | 92 | 93 | 94 | 95 | 96 | 97 | 98 | 99 | 100 |

a  1 is neither prime nor composite. Draw a star around it.

b  2 is a prime number. Circle it.

c  Lightly shade all the multiples of 2. They are composite numbers.

d  Put a circle around the next prime number: 3

e  Lightly shade all the multiples of 3. They are composite numbers.

f  Put a circle around the next prime number: 5

g  Lightly shade all the multiples of 5. They are composite numbers.

h  Find the **next** prime number. Circle it.

i  Lightly shade all its multiples.

j  Repeat Step h and Step i until you get to the end of the grid.

---

**2** a  The highest prime number on the grid is: ☐

b  True or false? All the prime numbers are odd. _____

c  True or false? More of the composite numbers are even than odd. _____

**3** All composite numbers are made by multiplying prime numbers. 6 is a composite number. It can be made by multiplying 2 prime numbers: 2 × 3. We can show it in a factor tree:

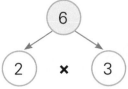

The prime factors of 6 are 2 and 3. So 6 = 2 × 3

*Prime factors are two or more prime numbers that are multiplied together to make a composite number.*

Fill in the gaps:

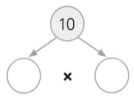 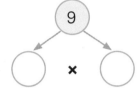 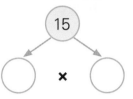

a  The prime factors of 10 are

_____.

b  The prime factors of 9 are

_____.

c  The prime factors of 15 are

_____.

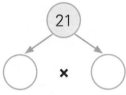 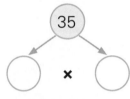 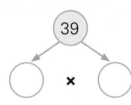

d  The prime factors of 21 are

_____.

e  The prime factors of 35 are

_____.

f  The prime factors of 39 are

_____.

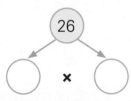 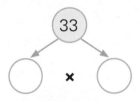 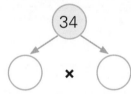

g  The prime factors of 26 are

_____.

h  The prime factors of 33 are

_____.

i  The prime factors of 34 are

_____.

**4** Draw factor trees for:

a  14

b  55

c  49

# Extended practice

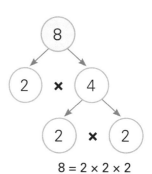

The prime factors of 8 are 2, 2 and 2. To show the prime factors of 8, we can write 2 × 2 × 2. We can also write $2^3$.

**1** Fill in the gaps.

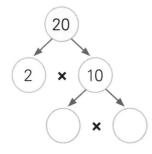
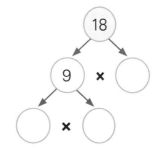

**a**  20 = 2 × 2 × _____
    20 = $2^{\square}$ × _____

**b**  18 = _____ × _____ × _____
    18 = $\square^{\square}$ × _____

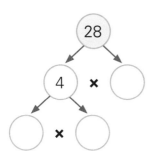
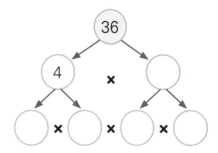

**c**  28 = _____ × _____ × _____
    28 = $\square^{\square}$ × _____

**d**  36 = _____ × _____ × _____ × _____
    36 = $\square^{\square}$ × $\square^{\square}$

---

**2** Draw factor trees to show the prime factors.

**a**  27

**b**  30

**c**  24

# UNIT 1: TOPIC 4
# Mental strategies for addition and subtraction

**Looking for short cuts**

Round numbers are easy to work with. For example, 287 − 98 = ?

We could say, 287 − 100 = 187.

We took away 2 too many, so we add 2 back to the answer. So, 287 − 98 = 189.

## Guided practice

**1** Use rounding for these subtractions. Fill in the gaps.

|   | Problem | Using rounding it becomes | Now I need to | Answer |
|---|---------|---------------------------|---------------|--------|
| a | 317 + 199 | 317 + 200 = 517 | take away 1 | 516 |
| b | 275 − 101 | 275 − 100 = 175 | take away another 1 |  |
| c | 527 + 302 | 527 +      = | add another    |  |
| d | 377 − 98 | 377 −      = |  |  |
| e | 249 + 249 |  |  |  |
| f | 938 − 206 |  |  |  |
| g | 1464 + 998 |  |  |  |

Splitting numbers can make addition easier.

For example, 160 + 830 = ?

Split (expand) the numbers: 100 + 60 + 800 + 30

Join the partners: 100 + 800 + 60 + 30 = 900 + 90 = 990

*Looking for sensible short cuts makes sense to me!*

**2** Split the numbers. Fill in the gaps.

|   | Problem | Expand the numbers | Join the partners | Answer |
|---|---------|-------------------|-------------------|--------|
| a | 370 + 520 | 300 + 70 + 500 + 20 | 300 + 500 + 70 + 20 | 890 |
| b | 2200 + 3600 | 2000 + 200 + 3000 + 600 | 2000 + 3000 + 200 + 600 |  |
| c | 342 + 236 | 300 + 40 + 2 + 200 + 30 + 6 |  |  |
| d | 471 + 228 |  |  |  |
| e | 743 + 426 |  |  |  |
| f | 865 + 734 |  |  |  |
| g | 4270 + 3220 |  |  |  |

## Independent practice

**1** Use the split strategy to solve these, or find your own sensible short cut.

a   147 + 232   _____

b   184 + 415   _____

c   747 + 551   _____

d   1552 + 732   _____

e   3267 + 642   _____

f   6564 + 4426   _____

**2** Use the rounding (compensation) strategy to solve these, or find another short cut.

a   745 − 299   _____

b   364 + 401   _____

c   276 + 598   _____

d   847 − 302   _____

e   958 − 190   _____

f   902 + 304   _____

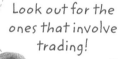

*Look out for the ones that involve trading!*

**3** Choose a strategy to solve these. Explain how you got each answer.

a   649 + 249 = _____

b   1253 − 199 = _____

c   1750 + 1750 = _____

d   14 578 − 410 = _____

**4** Rounding (and estimating) are useful strategies for mental calculations.
How do we know what number we should round to? Sometimes it is quite obvious: 69 rounds to 70 and 902 rounds to 900. Do we always round to the nearest ten or hundred?

Look at these facts and figures and decide how to round the numbers.
Explain how you have rounded them.

| | Number fact | Rounded number | I rounded this number to the nearest … |
|---|---|---|---|
| a | Australia has 812 972 kilometres of roads. | | |
| b | The Electricity Company of China employs 1 502 000 people. | | |
| c | The Mexican soccer player, Blanco, earned $2 943 702 in 2009. | | |
| d | The fastest speed recorded at the Indianapolis 500 car race was 299.3 km/h. | | |
| e | The fastest 100-metre sprint time for a woman is 10.49 seconds. | | |
| f | The US department store Walmart employs 2 100 000 people. | | |
| g | Each Australian eats an average of 17 L 600 mL of ice-cream a year. | | |
| h | The longest rail tunnel is in Switzerland. It is 57.1 km long. | | |
| i | The amount of money the movie *Avatar* made was $2 783 919 000. | | |
| j | Foreign tourists spend $29 127 000 000 a year in Australia. | | |

**5** A truck company is offering discounts. Use mental strategies to work out the new prices.

| Type | Basic | Deluxe | Premium |
|---|---|---|---|
| Was | $23 990 | $33 629 | $42 158 |
| Save | $1500 | $2139 | $3199 |
| New price | | | |

# Extended practice

**Are calculators always right?**

The answer is "yes", but only when they are used properly. Let's suppose Lee wants to add 249 523 + 248 614. Could she trust a calculator answer of 298 137?

It's easy to estimate the answer by rounding: 250 000 + 250 000 = 500 000.

So, the answer has to be close to 500 000 and not 300 000. The calculator answer was "wrong" because the wrong information was put into the calculator.

**1** Round and estimate to fill the gaps.

| | Problem | Round the numbers | Estimate the answer | Circle the likely answer |
|---|---|---|---|---|
| | 109 897 + 50 157 | 110 000 + 50 000 | 160 000 | 261 054 or (161 054) |
| a | 5189 – 2995 | | | 2194 or 3194 |
| b | 2958 + 6058 | | | 9016 or 8016 |
| c | 8215 – 3108 | | | 5907 or 5107 |
| d | 15 963 + 14 387 | | | 29 350 or 30 350 |
| e | 8954 – 3928 | | | 5026 or 4026 |
| f | 4568 + 4489 | | | 8057 or 9057 |
| g | 13 149 – 7908 | | | 6241 or 5241 |
| h | 124 963 + 98 358 | | | 223 321 or 213 321 |

**2** Estimate each answer, then check on a calculator. If the calculator answer is not close to your estimate, find out what went wrong.

| | Problem | Round the numbers | Estimate the answer | Calculator answer |
|---|---|---|---|---|
| | 6190 + 1880 | 6000 + 2000 | 8000 (Good estimate) | 8070 |
| a | 4155 + 2896 | | | |
| b | 9124 – 8123 | | | |
| c | 24 065 + 5103 | | | |
| d | 19 753 – 10 338 | | | |
| e | 101 582 + 49 268 | | | |
| f | 298 047 – 198 214 | | | |
| g | 1 089 274 + 1 099 583 | | | |
| h | 1 499 836 + 1 489 967 | | | |

# UNIT 1: TOPIC 5
## Written strategies for addition

**Everything in its place**

Errors in addition problems often result from not putting things in the right place.

For example, it is easy to see what went wrong when Jake tried to add 724 and 216:

```
      7 2 4                    7 ¹2 4
  +   2 1 6                +     2 1 6
    2 8 8 4                      9 4 0
```

If the digits were in the right columns, this addition problem would be easy to solve correctly.

## Guided practice

*Remember to trade where necessary.*

**1** Write these vertical algorithms. Look for a pattern in the answers.

a  85 + 1149

```
        8 5
  + 1 1 4 9
  _____
```

b  2029 + 316

c  2980 + 476

d  857 + 3710

e  873 + 4831 + 85

f  4759 + 87 + 832

## Independent practice

**1** Set out these addition problems vertically. Look for a pattern in the answers.

a   548 + 563

b   1325 + 897

c   1365 + 1968

d   3962 + 482

e   3290 + 869 + 1396

f   4378 + 1967 + 321

g   458 + 5379 + 1940

h   49 + 3721 + 578 + 4540

i   7357 + 768 + 64 + 745 + 1065

j   5396 + 546 + 54 + 3955 + 49

k   45 + 5348 + 543 + 43 + 345 + 4787

*Once you get the hang of it, you can add numbers of any size!*

**2** Make up a 5-line addition algorithm for which the answer is 99 999. Make sure each line of the algorithm has at least 3 digits.

**3** Did you know that in Australia there are more kilometres of unpaved roads than paved roads? However, in France, there are zero kilometres of unpaved roads.

a   Find the total length of roads for each country.

| Country | Length of paved roads (km) | Length of unpaved roads (km) | Total length of roads (km) |
|---|---|---|---|
| USA | 4 165 110 | 2 265 256 | |
| India | 1 603 705 | 1 779 639 | |
| China | 1 515 797 | 354 864 | |
| France | 951 220 | 0 | |
| Japan | 925 000 | 258 000 | |
| Spain | 659 629 | 6 663 | |
| Canada | 415 600 | 626 700 | |
| Australia | 336 962 | 473 679 | |
| Brazil | 96 353 | 1 655 515 | |

b   Apart from Australia, which other countries have more kilometres of unpaved roads than paved roads? _____

c   The total of paved roads in which two countries is 5 680 907?
_____

d   The unpaved roads of which two countries are closest to 1 million kilometres?
_____

# Extended practice

## Foreign students

Use the table to complete these activities.

| Countries with the highest numbers of foreign students | | |
|---|---|---|
| | USA | 595 874 |
| | UK | 351 470 |
| | France | 246 612 |
| | Australia | 211 526 |
| | Germany | 206 875 |

**1** Write a vertical algorithm and find the total number of foreign students in the USA and UK.

**2** Some people use a calculator to check an answer. Find the total number of foreign students in Australia and the UK by writing an algorithm. Then check the answer with a calculator.

**3** Imagine you wanted to find the total number of foreign students in France and Germany and in your written algorithm the answer is 453 487. The calculator check gives the answer 453 397. Which answer would you trust? Check to find the correct answer.

## Working with trillions

**4** How much of our world is covered by oceans? Use the table to find out.

| Pacific Ocean | 155 557 000 000 km² |
|---|---|
| Atlantic Ocean | 76 762 000 000 km² |
| Indian Ocean | 68 656 000 000 km² |
| Southern Ocean | 20 327 000 000 km² |
| Arctic Ocean | 14 056 000 000 km² |

Answer: _____

# UNIT 1: TOPIC 6
## Written strategies for subtraction

**Trading**

To make 3465 − 1329 easier to solve, we need to trade. It's like rewriting the number on the top line:

3000 + 400 + 60 + 5 is the same as
3000 + 400 + 50 + 15

There aren't enough ones. Trade a ten. That leaves 5 tens.

| Th | H | T | O |
|---|---|---|---|
| 3 | 4 | ⁵6̸ | ¹5 |
| − 1 | 3 | 2 | 9 |
| 2 | 1 | 3 | 6 |

Now there are 10 + 5 ones = 15

## Guided practice

**1**

a)
| H | T | O |
|---|---|---|
| 3 | ⁶7̸ | ¹1 |
| − 1 | 4 | 2 |
|   |   |   |

b)
| H | T | O |
|---|---|---|
| 8 | 5̸ | 4 |
| − 5 | 2 | 8 |
|   |   |   |

c)
| Th | H | T | O |
|---|---|---|---|
| 4 | 2 | 3 | 6 |
| − 2 | 0 | 2 | 8 |
|   |   |   |   |

d)
| Th | H | T | O |
|---|---|---|---|
| 6 | 2 | 7 | 3 |
| − 4 | 1 | 5 | 4 |
|   |   |   |   |

**2**

a)
| H | T | O |
|---|---|---|
| 8 | 3 | 6 |
| − 2 | 4 | 7 |
|   |   |   |

b)
| H | T | O |
|---|---|---|
| 5 | 3 | 8 |
| − 3 | 3 | 9 |
|   |   |   |

c)
| Th | H | T | O |
|---|---|---|---|
| 5 | 6 | 2 | 0 |
| − 3 | 4 | 7 | 1 |
|   |   |   |   |

d)
| Th | H | T | O |
|---|---|---|---|
| 4 | 3 | 8 | 4 |
| − 2 | 3 | 9 | 9 |
|   |   |   |   |

e)
| Tth | Th | H | T | O |
|---|---|---|---|---|
| 5 | 3 | 6 | 1 | 5 |
| − 4 | 3 | 6 | 2 | 7 |
|   |   |   |   |   |

f)
| Tth | Th | H | T | O |
|---|---|---|---|---|
| 2 | 3 | 5 | 9 | 8 |
| − 1 | 4 | 6 | 9 | 9 |
|   |   |   |   |   |

# Independent practice

**1** Complete these. Look for a pattern in the answers.

a)   9 2 5 4 5
   − 3 8 2 2 4

b)   8 4 0 4 7
   − 1 8 6 1 5

c)   9 1 3 6 0
   − 1 4 8 1 7

d)   9 2 9 7 2
   −   5 3 1 8

e)   9 9 9 5 3
   −   1 1 8 8

f)   8 8 2 2 5
   − 3 1 4 3 6

g)   7 7 7 5 6
   − 3 2 0 7 8

h)   4 6 6 3 5
   − 1 2 0 6 8

i)   9 7 7 4 6
   − 7 4 2 9 0

j)   2 1 2 1 2
   −   8 8 6 7

---

**2** Use the following digits once each to make the largest and smallest numbers possible. Then find the difference between them.

8   3   7   2   4   5   9

Working-out space

**3** You can also trade across more than one column.

> Remember to trade across one column at a time.

More ones are needed ...
... but there are no tens

| Th | H | T | O |
|----|---|---|---|
| 3  | 4 | 0 | ⑤ |
| – 1 | 3 | 2 | 9 |

... so trade from the HUNDREDS to the TENS first.

Trade a hundred. That leaves 3 hundreds.

| Th | H | T | O |
|----|---|---|---|
| 3  | ³4̷ | ¹0 | 5 |
| – 1 | 3 | 2 | 9 |

Now there are 10 tens.

Trade a ten. That leaves 9 tens.

| Th | H | T | O |
|----|---|---|---|
| 3  | ³4̷ | ⁹1̷0̷ | ¹5 |
| – 1 | 3 | 2 | 9 |
|    | 2 | 0 | 7 | 6 |

Now there are 15 ones.

a)  1 2 7 0 4
   –    9 4 3 6

b)  2 5 0 1 2
   –  1 2 3 9 3

c)  4 0 3 0 4
   –  1 7 6 4 8

d)  5 0 4 0 8
   –  1 5 8 2 9

e)  5 0 5 2 0 5
   –  1 2 9 4 2 8

f)  9 0 3 4 0 5
   –  2 2 7 3 3 7

g)  2 0 0 8
   –  1 2 5 9

h)  1 6 0 0 2
   –  1 2 3 5 3

i)  7 4 0 0 5 8
   –  4 2 0 0 0 4

j)  1 0 0 0 0 0
   –    3 4 3 7 8

# Extended practice

## Checking subtraction by using addition

**1** One way to check a subtraction answer is by addition. For example, 100 − 75 is 25. We can be sure of this by adding 25 to 75.

**a** Does 317 418 − 123 783 = **193 635** or **193 335**? Check by adding, and then complete the subtraction.

```
    1 9 3 6 3 5         1 9 3 3 3 5         3 1 7 4 1 8
+   1 2 3 7 8 3     +   1 2 3 7 8 3     −   1 2 3 7 8 3
  _____      _____      _____

  _____      _____      _____
```

**b** Does 326 175 − 199 879 = **126 296** or **126 206**? Check by adding, and then complete the subtraction.

```
    1 2 6 2 9 6         1 2 6 2 0 6         3 2 6 1 7 5
+   1 9 9 8 7 9     +   1 9 9 8 7 9     −   1 9 9 8 7 9
  _____      _____      _____

  _____      _____      _____
```

**c** Does 350 044 − 158 254 = **192 890** or **191 790**? Check by adding, and then complete the subtraction.

```
    1 9 2 8 9 0         1 9 1 7 9 0         3 5 0 0 4 4
+   1 5 8 2 5 4     +   1 5 8 2 5 4     −   1 5 8 2 5 4
  _____      _____      _____

  _____      _____      _____
```

**2** Here is a subtraction and addition trick using 3 different digits, such as 3, 6 and 2 or 3, 4 and 5. Try with other sets of 3 digits here and on another sheet of paper.

| | | | | | | | | | | | |
|---|---|---|---|---|---|---|---|---|---|---|---|
| Make the largest number | | 6 | 3 | 2 | | | 5 | 4 | 3 | | |
| Make the smallest number | − | 2 | 3 | 6 | | − | 3 | 4 | 5 | | |
| Subtract | | 3 | 9 | 6 | OR | | 1 | 9 | 8 | OR | |
| Reverse the number | + | 6 | 9 | 3 | | + | 8 | 9 | 1 | | |
| Add | 1 | 0 | 8 | 9 | | 1 | 0 | 8 | 9 | | |

**a** Use other sets of 3 different digits to prove that the answer is always the same.

**b** What happens if two of the three digits are the same?

_____

# UNIT 1: TOPIC 7
## Mental strategies for multiplication and division

**The ten trick**

Multiplying by 10 in your head is easy – but you don't just add a zero!

14, 82, $1.50, 7, 35, 725  × 10  →  140, 820, $15.00, 70, 350, 7250

*If we just added a zero to multiply $1.50 by 10, the answer would be $1.500 – and that's not correct!*

## Guided practice

**1** When you **multiply** by 10, the digits move one place bigger (to the left) and the zero fills the space. When you multiply by 100 they move two places, and so on. Complete the grid.

| × | 10 | 100 | 1000 | 10 000 |
|---|---|---|---|---|
| 37 | 370 | 3700 | 37 000 | 370 000 |
| a  29 | | | | |
| b  124 | | | | |
| c  638 | | | | |
| d  $1.25 | | | | |
| e  750 | | | | |

**2** When you **divide** by 10, the digits move one place smaller (to the right). Use decimals if necessary.

| ÷ | 10 | Write the multiplication fact partner |
|---|---|---|
| 120 | 12 | 12 × 10 = 120 |
| 45 | 4.5 | 4.5 × 10 = 45 |
| a  370 | | |
| b  4700 | | |
| c  2000 | | |
| d  $22.50 | | |
| e  54 | | |

**3** When you **divide** by 100, the digits move two places to the right. Use decimals if necessary.

| ÷ | 100 | Write the multiplication fact partner |
|---|---|---|
| 500 | 5 | 5 × 100 = 500 |
| $275 | $2.75 | $2.75 × 100 = $275 |
| a  700 | | |
| b  $495 | | |
| c  5000 | | |
| d  12 000 | | |
| e  8750 | | |

# Independent practice

**1** Multiplying by multiples of 10 using doubling strategies.

| × | | 10 | 20 [double] | 40 [double again] | 80 [double again] |
|---|---|---|---|---|---|
| | 13 | 130 | 260 | 520 | 1040 |
| a | 12 | | | | |
| b | 15 | | | | |
| c | 22 | | | | |
| d | 25 | | | | |
| e | 50 | | | | |

**2** Dividing by multiples of 10 using halving strategies.

| ÷ | | ÷ 10 | ÷ 20 [halve it] | ÷ 40 [halve again] | ÷ 80 [halve again] |
|---|---|---|---|---|---|
| | 800 | 80 | 40 | 20 | 10 |
| a | 400 | | | | |
| b | 2000 | | | | |
| c | 480 | | | | |
| d | 10 000 | | | | |
| e | 8800 | | | | |

**3** Multiplying large numbers by 5.

| × 5 | | First multiply by 10 | Then halve it | Multiplication fact |
|---|---|---|---|---|
| | 84 | 840 | 420 | 84 × 5 = 420 |
| a | 24 | | | |
| b | 68 | | | |
| c | 120 | | | |
| d | 500 | | | |
| e | 1240 | | | |

*The secret is to find a short cut that works for YOU!*

**4** Dividing large numbers by 5.

| ÷ 5 | | First divide by 10 | Then double it | Division fact |
|---|---|---|---|---|
| | 160 | 16 | 32 | 160 ÷ 5 = 32 |
| a | 420 | | | |
| b | 350 | | | |
| c | 520 | | | |
| d | 900 | | | |
| e | 1200 | | | |

**5** Multiplying by splitting the multiple of 10.

25 × 30 is the same as 25 × 10 three times, so you can find the answer by splitting 30 into 3 tens.

| | × 30 | First × 10 | Then × 3 | Multiplication fact |
|---|---|---|---|---|
| | 25 | 250 | 750 | 25 × 30 = 750 |
| a | 15 | | | |
| b | 22 | | | |
| c | 33 | | | |
| d | 150 | | | |
| e | 230 | | | |

**6** Sometimes it's easier to split the multiple of 10 differently. 25 × 30 is also the same as finding 25 × 3, ten times.

| | × 30 | First × 3 | Then × 10 | Multiplication fact |
|---|---|---|---|---|
| | 25 | 75 | 750 | 25 × 30 = 750 |
| a | 15 | | | |
| b | 22 | | | |
| c | 33 | | | |
| d | 150 | | | |
| e | 230 | | | |

**7** Use your choice of strategy. Be ready to explain how you got the answer.

a  15 × 40  _____    b  22 × 40  _____

c  25 × 50  _____    d  34 × 50  _____

e  14 × 60  _____    f  125 × 40  _____

g  15 × 80  _____    h  72 × 20  _____

i  19 × 30  _____    j  $1.20 × 60  _____

k  $2.25 × 40  _____    l  832 ÷ 2  _____

m  832 ÷ 4  _____    n  248 ÷ 4  _____

**8** Sam has a coin collection with 437 twenty-cent coins in it. Use a mental strategy to find out how much money Sam has.

Answer: _____

# Extended practice

**1** You can use the split strategy to multiply 24 by 15. Split it into 24 × 10 and 24 × 5.

|   | × 15 | × 10 | Halve it to find × 5 | Add the two answers | Multiplication fact |
|---|------|------|----------------------|---------------------|---------------------|
|   | 24   | 240  | 120                  | 240 + 120 = 360     | 24 × 15 = 360       |
| a | 12   |      |                      |                     |                     |
| b | 32   |      |                      |                     |                     |
| c | 41   |      |                      |                     |                     |
| d | 86   |      |                      |                     |                     |
| e | 422  |      |                      |                     |                     |

**2** Here is a mental strategy for multiplying by 13.

|   | × 13 | Number × 10 | Number × 3 | Add the two answers | Multiplication fact |
|---|------|-------------|------------|---------------------|---------------------|
|   | 22   | 220         | 66         | 220 + 66 = 286      | 22 × 13 = 286       |
| a | 15   |             |            |                     |                     |
| b | 12   |             |            |                     |                     |
| c | 23   |             |            |                     |                     |
| d | 31   |             |            |                     |                     |
| e | 25   |             |            |                     |                     |

**3** Choose a strategy to solve these problems. Be ready to explain how you got the answers.

a  25 × 100 _____   b  315 × 20 _____

c  80 ÷ 20 _____   d  900 ÷ 30 _____

e  22 × 400 _____   f  $36 ÷ 20 _____

g  $3.40 × 20 _____   h  $36 ÷ 40 _____

**4** When you are resting, your heart beats about once every second. About how many times does your heart beat in

a  1 minute? _____   b  1 hour? _____

# UNIT 1: TOPIC 8
# Written strategies for multiplication

**Extended and short multiplication**

You can write the multiplication for 43 × 5 in either its long or short form.

**Extended multiplication**

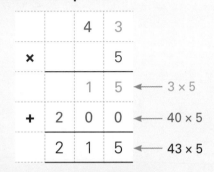

**Short multiplication**

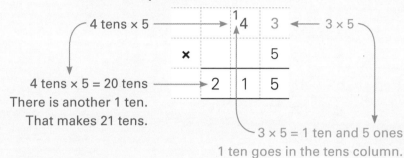

## Guided practice

**1** Extended and short multiplication.

a)
```
      5 4
  ×     3
  -----
      1 2
  +   0
  -----
```

```
     ¹5 4
  ×    3
  -----
```

b)
```
      6 5
  ×     5
  -----
  +   0
  -----
```

```
     ²6 5
  ×    5
  -----
```

**2**

a)
```
    1 2 7
  ×     4
  -----
```

b)
```
    3 2 7
  ×     3
  -----
```

c)
```
    3 1 5
  ×     2
  -----
```

You could do these as extended multiplication on a separate piece of paper.

d)
```
    2 2 9
  ×     4
  -----
```

e)
```
    1 6 3 8
  ×       5
  -----
```

f)
```
    1 3 4 5
  ×       7
  -----
```

g)
```
    2 1 2 8
  ×       4
  -----
```

h)
```
    1 4 5 7
  ×       5
  -----
```

i)
```
    1 5 0 7
  ×       6
  -----
```

## Independent practice

Multiplying by a multiple of 10 is no harder than multiplying by a single-digit number, as long as you remember the "10 trick". For example, what is 43 × 20?

It's the 10 trick, so everything moves over one place.

```
    4 3
×   2 0
-------
  8 6 0
```

1. Put a ZERO as a space-filler.
2. Then just multiply × 2.

**1**

a)
```
    3 5
×   2 0
-------
      0
```

b)
```
    2 7
×   2 0
-------
      0
```

c)
```
    3 6
×   3 0
-------
```

d)
```
    4 6
×   4 0
-------
```

e)
```
    6 7
×   3 0
-------
```

f)
```
    3 4
×   6 0
-------
```

Remember the zero — it moves everything over one place.

g)
```
  1 5 2
×   2 0
-------
```

h)
```
  2 4 6
×   4 0
-------
```

i)
```
  1 8 3
×   6 0
-------
```

j)
```
1 3 8 2
×   4 0
-------
```

k)
```
2 6 5 8
×   7 0
-------
```

l)
```
2 6 0 9
×   8 0
-------
```

**2** Thirty students go on a school camp. It costs $146 for each student. What is the total cost?

Working-out space

**3** The crystal in a quartz watch vibrates 32 768 times a second. How many times does it vibrate in one minute?

OXFORD UNIVERSITY PRESS

To multiply by two digits, split the number you are multiplying by.

**What is 36 × 25?**
There are two multiplications.

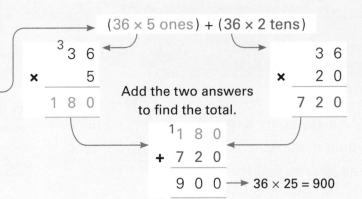

$36 \times 25 = 900$

**4**  a  What is 24 × 23?   b  What is 23 × 35?   c  What is 35 × 28?

Make two multiplications.

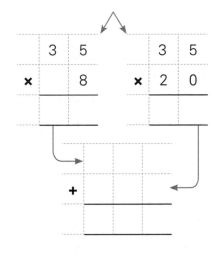

24 × 23 = ☐   23 × 35 = ☐   35 × 28 = ☐

**5**  a  What is 37 × 24?   b  What is 39 × 27?   c  What is 42 × 26?

Make two multiplications.

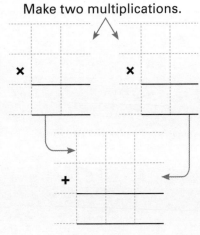 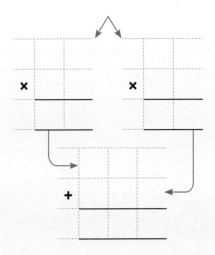 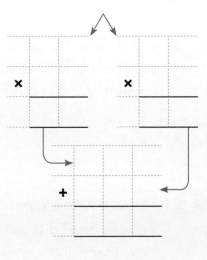

37 × 24 = ☐   39 × 27 = ☐   42 × 26 = ☐

# Extended practice

At the top of page 36, you looked at a way of multiplying 36 × 25. You can make this shorter by putting both multiplications into one algorithm: (36 × 5) + (36 × 20).

**What is 36 × 25?**

```
           3 6
       ×   2 5
36 × 5 →   1 8 0
36 × 20 → +7 2 0
           9 0 0
```

Don't forget to put a zero as a space-filler.

**1**

a
```
    2 9
  × 2 5
         ← 29 × 5
  +    0 ← 29 × 20
  _____
```

b
```
    4 2
  × 2 7
  _____
  +
  _____
```

c
```
    3 9
  × 1 9
  _____
  +
  _____
```

d
```
    3 3
  × 4 3
  _____
  +
  _____
```

e
```
    7 5
  × 1 5
  _____
  +
  _____
```

f
```
    6 4
  × 3 7
  _____
  +
  _____
```

g
```
    1 2 3
  ×   2 6
  _____
  +
  _____
```

h
```
    2 0 7
  ×   5 4
  _____
  +
  _____
```

i
```
    3 9 6
  ×   4 7
  _____
  +
  _____
```

**2** Use the working-out space to answer these.

a At Mount Waialeale in Hawaii, it rains 335 days a year. If you lived there for 35 years, how many rainy days would you have?

b Amy sneezed 2700 times a day for two years. How many times did she sneeze in the month of January?

Working-out space

# UNIT 1: TOPIC 9
# Written strategies for division

**Two ways of writing division problems**

150 divided by 2 can be written as 150 ÷ 2 or as 2)‾150‾

If you need to do written working for a problem such as 135 ÷ 3, do it like this:

| There aren't enough hundreds to make groups of 3. We start with 13 tens. 13 tens split into groups of 3 = 4 r1 | Trade the ten for 10 ones. This makes 15 ones. 15 split into groups of 3 = 5 |
|---|---|
|     4<br>3 )1 3 5 |     4 5<br>3 )1 3 ¹5 |

*Keep the answer digits in the right columns.*

## Guided practice

**1** Complete these extended and short divisions.

a   4)2 7³6  (with 6 above)
b   2)8 8 4
c   7)6 5 8

d   4)4 4 0
e   2)8 6 4 2
f   3)3 6 0 3

g   4)3 7 3 6
h   5)2 4 2 0 5
i   7)3 0 2 5 4

j   6)2 5 9 3 2
k   8)9 8 7 4 4
l   9)4 8 8 8 9 8

wrong     right

H T O     H T O
  4 5        4 5
3)135     3)135

**2** Rewrite using the )‾ symbol and solve.

a   1075 ÷ 5
b   1746 ÷ 3
c   2148 ÷ 6

d   5372 ÷ 2
e   2636 ÷ 4
f   2436 ÷ 7

## Independent practice

Here are two ways of showing a remainder at the end of a division problem:

$$35 \div 2 = 17 \text{ r}1 \quad \text{or} \quad 35 \div 2 = 17\frac{1}{2}$$

**1** Write the remainder in two ways.

a  14 ÷ 3 _____    b  47 ÷ 5 _____

c  39 ÷ 4 _____    d  65 ÷ 8 _____

e  77 ÷ 9 _____    f  61 ÷ 7 _____

g  84 ÷ 9 _____    h  58 ÷ 6 _____

**2** Complete each algorithm, showing the remainder in two ways.

a  4 ) 4 6 7        4 ) 4 6 7        b  3 ) 2 7 2        3 ) 2 7 2

c  6 ) 1 9 7        6 ) 1 9 7        d  5 ) 7 4 2        5 ) 7 4 2

e  3 ) 2 5 7 5      3 ) 2 5 7 5      f  9 ) 1 6 8 4      9 ) 1 6 8 4

g  6 ) 4 1 6 5      6 ) 4 1 6 5      h  7 ) 2 3 1 9      7 ) 2 3 1 9

**3** In a real-world division problem, we have to decide what to do with a remainder. Do we leave it as a remainder, or divide it into fractions? Give a real-world answer to each of these problems. Be ready to justify your answer.

a  Two children share a bag of 125 marbles. How many do they each get?

_____

b  Two children share 15 donuts. How many do they each get?

_____

If two people shared $25, we would not leave the remainder as $1, nor would we call it $\frac{1}{2}$ a dollar. We would use a decimal: $25 ÷ 2 = $12.50

To write an algorithm, we need to put a decimal point and show "zero cents".

$$\begin{array}{r} \$\phantom{0}12.50 \\ 2\overline{)25.00} \end{array}$$

**4** Put in the decimal point and the zeros to complete these.

a  $2\overline{)53.00}$     b  $4\overline{)74}$     c  $8\overline{)92}$

d  $4\overline{)73}$     e  $8\overline{)132}$     f  $6\overline{)129}$

---

**5** We can use decimals for other remainders.

For example, if Than scores $\frac{17}{20}$, $\frac{18}{20}$, $\frac{19}{20}$ and $\frac{15}{20}$ in four tests, his average score is the total (69) ÷ 4 = 17 r1, or $17\frac{1}{4}$ or $4\overline{)6^29.^10^20}$ giving 17.25

a  $4\overline{)595.00}$     b  $5\overline{)628}$     c  $8\overline{)506}$

d  $5\overline{)684}$     e  $4\overline{)1347}$     f  $8\overline{)9852}$

g  $6\overline{)17193}$     h  $5\overline{)11598}$     i  $8\overline{)52186}$

---

**6** Solve the following problems. Think of the most appropriate way to deal with the remainders.

a  145 marbles are divided between four people. How many do they each have?

b  Four people share a prize of $145. How much does each person receive?

# Extended practice

Sometimes a decimal remainder goes further than two decimal places. We can choose to stop after a certain number of places. For example, if Than scores 18 out of 20, then 15, then 19, his average score is 52 divided by 3:

Stop after 2 decimal places = 17.33

```
    1 7.3 3 3 3 3
3 ) 5 2.0 0 0 0 0 0
```

**1** Show the remainder as a decimal.

*Stop after two decimal places.*

a  3 ) 8 7 4     b  4 ) 4 9 7     c  7 ) 5 8 2

d  6 ) 2 5 4     e  9 ) 7 2 4     f  3 ) 5 4 8 5

g  5 ) 1 7 4 3   h  7 ) 8 5 8 3   i  4 ) 4 5 9 7 9

j  6 ) 8 5 9 2 8   k  5 ) 2 5 4 7 6   l  9 ) 9 7 2 6 5

**2** Write the correct digit in each gap.

a  4 ) ☐ 3 2 = 58

b  7 ) 3 ☐ 9 = 4 ☐7

c  ☐ ) 1 9 1 3 = 478.25

d  3 ☐ 6 ÷ 8 = 42

e  $726.50 ÷ ☐ = $363.25

f  1837.☐ ÷ 3 = 612.5

g  2643.75 ÷ ☐ = 528.75

**3** The geyser Old Faithful in the United States erupts 7300 times in a regular year (not a leap year). How many times does it erupt each **day**?

Working-out space

# UNIT 1: TOPIC 10
## Integers

If you ask a seven-year-old, "What is 5 – 8?" they will probably answer, "You can't take 8 away from 5." However, a calculator would give the answer:

5 – 8 = –3

Integers are whole numbers. They can be *positive* (greater than zero) or *negative* (less than zero). For –3, we say *negative* 3.

## Guided practice

**1** Fill in the gaps on this number line.

**2** The red dot is at zero. Draw these shapes on the number line:

a   a blue dot at –3
b   a black dot at 2
c   a triangle at –1
d   a square at 4
e   a star at –5

**3** The value of the highest number for the shapes in question 2 is 4. Write the number values of the shapes from **lowest** to **highest**.

_____

**4** 1 > –3 means that 1 has a greater value than negative 3. Write **True** or **False** for these statements:

*The > sign means bigger than and < means less than.*

a   5 > 0   _____
b   0 < –1   _____
c   2 > –4   _____
d   –2 > –1   _____
e   –4 < 0   _____
f   5 = –5   _____
g   3 < –4   _____
h   –5 > –10   _____

# Independent practice

A number line can show operations such as 2 + 4 or 1 − 3.

**Increase 2 by 4.**

Number sentence: 2 + 4 = 6

**Decrease 1 by 3.**

Number sentence: 1 − 3 = −2

**1** Show these operations on the number lines. Write the number sentence for each operation.

a  Increase −2 by 4.
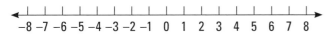
Number sentence: _____

b  Decrease 2 by 3.
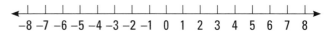
Number sentence: _____

c  Decrease 4 by 7.

Number sentence: _____

d  Increase −6 by 5.
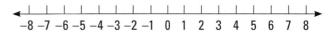
Number sentence: _____

e  Decrease −3 by 5.

Number sentence: _____

f  Increase −8 by 8.

Number sentence: _____

g  Increase −8 by 10.
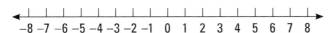
Number sentence: _____

h  Decrease 7 by 11.

Number sentence: _____

i  Increase −7 by 15.

Number sentence: _____

j  Decrease 6 by 13.

Number sentence: _____

**2** Write what a calculator would show if you pressed the following:

a  4 − 5 = _____

b  15 − 16 = _____

c  4 − 8 = _____

d  7 − 12 = _____

e  10 − 20 = _____

f  40 − 100 = _____

**3** Find the counting number for these number lines, then fill in the missing numbers.

a  −60 ☐ ☐ ☐ ☐ ☐ ☐ ☐ ☐ ☐ ☐  50

b  −25 ☐ ☐ ☐ ☐ ☐ ☐ ☐ ☐ ☐ ☐  30

c  −28 ☐ ☐ ☐ ☐ ☐ ☐ ☐ ☐ ☐ ☐  16

d  −35 ☐ ☐ ☐ ☐ ☐ ☐ ☐ ☐ ☐ ☐  42

e  −63 ☐ ☐ ☐ ☐ ☐ ☐ ☐ ☐ ☐ ☐  36

**4** The table shows the 4 am temperatures over one week in a mountain range.

| Saturday | −1°C |
| --- | --- |
| Sunday | 1°C |
| Monday | −2°C |
| Tuesday | 2°C |
| Wednesday | 0°C |
| Thursday | −4°C |
| Friday | −3°C |

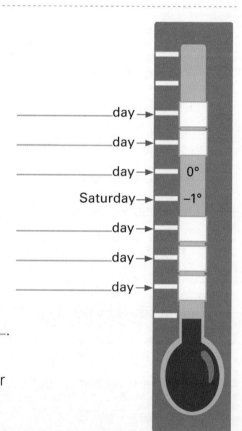

a  Write the temperatures on the thermometer.

b  Write the correct day next to each temperature.

c  The coldest day was _____.

d  The coldest temperature was _____° colder than the warmest.

**5** Each of these letters represents the value of a number. From the information given, plot the letters in the correct place on the number line.

−5  −4  −3  −2  −1  0  1  2  3  4  5
☐ ☐ ☐ ☐ ☐ ☐ ☐ ☐ ☐ ☐ ☐

- H > 3 but < 5
- S > H
- M < A
- T < 0 but > −2
- A < −3 but > −5

# Extended practice

**1** Use the graph to answer these questions.

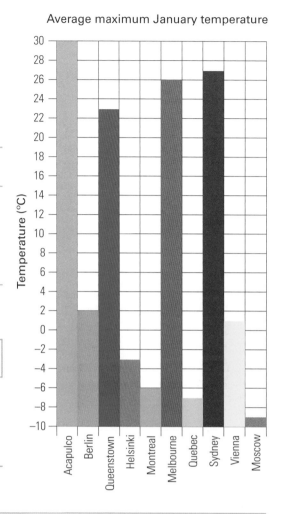

Average maximum January temperature

a  Which cities have negative average January temperatures?

_____

_____

b  In which city is the average temperature 5°C higher than Helsinki?

_____

c  If the average temperature in Vienna were to fall by 6°C, what would the new average temperature be? ☐

d  Which three pairs of cities have a temperature difference of 33°C?

_____

_____

---

**2** Sometimes banks allow people to withdraw money even if there is not enough money in their account. This is like a loan. When this happens, the person has a negative amount of money. A bank statement is one way to check how much is in the account. Fill in the balances.

| INTERNATIONAL BIG BANK | | | |
|---|---|---|---|
| Date | Paid in $ | Paid out $ | Balance $ |
| 3 May | 100 | | 100 |
| 4 May | | 120 | |
| 9 May | 30 | | |
| 14 May | | 50 | |
| 31 May | 45 | | |

---

**3** People often use credit cards for shopping. Until a credit card is used, the amount owing on it is neither negative nor positive. The balance is zero.

a  If Tran uses a credit card to pay a $100 bill, what is the balance? _____

b  At the end of the month, Tran can choose to pay off some or all of his negative balance. If he chooses to pay back $10, he would still owe **more** than $90. Why do you think this is?

_____

# UNIT 1: TOPIC 11
# Exponents and square roots

**Exponents**

We often look for shortcuts in mathematics. A shortcut for 3 + 3 + 3 + 3 + 3 is 3 × 5 = 15.

We can also use **exponents** as shortcuts. A short way of writing 3 × 3 is $3^2$. The base number is 3. The exponent is 2. It can also be called the index or power. The exponent tells us to use the base number (3) in a multiplication two times. So, $3^2 = 3 \times 3 = 9$.

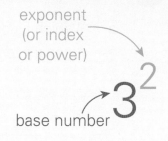

## Guided practice

**1** Write each multiplication as a base number and exponent. Remember to write the exponent smaller and higher than the base number.

| | Multiplication | Base number and exponent |
|---|---|---|
| | 3 × 3 × 3 | $3^3$ |
| a | 2 × 2 × 2 × 2 × 2 | |
| b | 4 × 4 × 4 | |
| c | 8 × 8 × 8 × 8 | |
| d | 5 × 5 × 5 × 5 × 5 | |
| e | 7 × 7 × 7 × 7 × 7 × 7 | |
| f | 10 × 10 × 10 × 10 | |

**2** Fill in the gaps in the table.

| | Base number and exponent | Number of times the base number is used in a multiplication | Multiplication | Value of the number |
|---|---|---|---|---|
| | $4^2$ | two times | 4 × 4 | 16 |
| a | $3^3$ | three times | | |
| b | $2^4$ | | | |
| c | $5^3$ | | | |
| d | $6^2$ | | | |
| e | $9^2$ | | | |
| f | $10^3$ | | | |

*For $3^3$, we can say 3 to the third power, 3 to the power of three or 3 cubed.*

## Square roots

To understand what is meant by the square root of a number, we need to look at square numbers.

Four squared can be written as $4^2$. The diagram on the right shows what it looks like.

$4^2 = 4 \times 4 = 16$

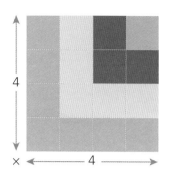

A square root goes the opposite way. The symbol for square root is $\sqrt{}$.

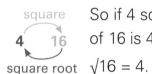

So if 4 squared is 16, the square root of 16 is 4. We can write it like this: $\sqrt{16} = 4$.

## Guided practice

**3** Find the square root of these numbers. Remember to ask the question: what number multiplied by itself makes the number?

| | Starting number | What number multiplied by itself makes the number? | Square root of the starting number | Number fact |
|---|---|---|---|---|
| | 16 | 4 × 4 = 16 | 4 | $\sqrt{16} = 4$ |
| a | 4 | | | |
| b | 36 | | | |
| c | 9 | | | |
| d | 64 | | | |

The starting numbers in question 3 were square numbers. If the starting number is not a square number, then we give the approximate the square root.

Find the approximate square roots of these numbers.

| | Starting number | Which two square numbers is it between? | What are their square roots? | The square number is between |
|---|---|---|---|---|
| | 7 | 4 and 9 | $\sqrt{4} = 2$ and $\sqrt{9} = 3$ | 2 and 3 |
| a | 10 | | | |
| b | 42 | | | |
| c | 20 | | | |
| d | 52 | | | |

# Independent practice

Base numbers with exponents can look small, such as $2^7$. However, when you expand the number the value can be quite large. For example, the value of $2^7$ is greater than 100.

**1** Find the value of these numbers. Begin by expanding the number. You may need a calculator for some.

   a  $2^7 = 2 \times 2 \times 2 \times 2 \times 2 \times 2 \times 2 = $ _____

   b  $5^5 = $ _____

   c  $3^6 = $ _____

   d  $4^5 = $ _____

   e  $7^4 = $ _____

**2** Circle the number with the greater value in each pair.   a  $9^4$ or $8^5$   b  $5^3$ or $3^5$

**3** Find the value of the exponent.

   a  5 to the power of _____ = 15 625

   b  10 to the power of _____ = 1 million

**4** Find the approximate square root, then the actual square root (to two decimal places). You will need a calculator with a square root function for the actual square root.

| | Starting number | The approximate square root is between | Actual square root (to two decimal places) | Number fact |
|---|---|---|---|---|
| | 5 | 2 and 3 | 2.24 | $\sqrt{5} = 2.24$ |
| a | 40 | | | |
| b | 14 | | | |
| c | 30 | | | |
| d | 99 | | | |

# Extended practice

Sometimes you will see $2^5$ written like this: 2^5. This can be useful if you are using a computer. The ^ symbol is usually above the 6 key.

**1** Find the value of the following.

a  5^2 = _____   b  3^4 = _____

c  10^4 = _____   d  1^10 = _____

## Negative exponents

A base number can have a negative exponent, like $2^{-2}$. Negative is the opposite of positive. A positive exponent involves multiplication. For example, $2^2 = 2 \times 2 = 4$.
The opposite of multiplication is division. A negative exponent involves division.
A negative exponent tells us how many times to divide 1 by the base number.
$2^{-2}$ tells us to **divide 1** by the base number (2) and then divide by 2 a second time.
First time: Divide 1 by 2 = $\frac{1}{2}$ or 0.5.
Second time: Divide $\frac{1}{2} \div 2 = \frac{1}{4}$ or 0.25.
So, $2^{-2} = 1 \div 2 \div 2 = 0.25$.

**2** Find the negative exponents. You may need a calculator for some.

a  $8^{-1} = 1 \div 8 =$ _____   b  $8^{-2} = 1 \div 8 \div 8 =$ _____

c  $4^{-1} =$ _____   d  $4^{-2} =$ _____

e  $10^{-2} =$ _____   f  $10^{-3} =$ _____

**3** In question 2, we started with 1 and then divided. A different way of looking at positive exponents it is to start at 1 and then multiply. For example, $3^2 = 1 \times 3 \times 3 = 9$. Find the value of these by expanding in the same way.

a  $6^3$   b  $4^4$

**4** What if the exponent is 1? Try with various numbers. Write a sentence about what happens to the base number when the exponent is 1.

# UNIT 2: TOPIC 1
# Fractions

**A fraction is a part of something**

A fraction can be part of a whole thing:

Three-eighths of the circle are shaded.

The frog is $\frac{3}{4}$ of the way along the line.

A fraction can also be part of a whole group, or quantity:

A quarter of the beads are red.

## Guided practice

**1** Write these fractions in words and numbers.

| a | What fraction is red? | b | What fraction is white? | c | What fraction is blue? | d | What fraction is green? |

one-_____     _____     _____     _____

**2** Write these fractions.

a  What fraction is going towards Melbourne?

b  Shade $\frac{1}{3}$ of the stars.

**3** The diamond is $\frac{1}{10}$ of the way along the number line.

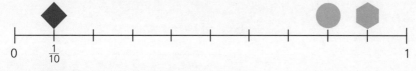

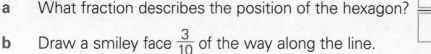

a  What fraction describes the position of the hexagon? _____

b  Draw a smiley face $\frac{3}{10}$ of the way along the line.

c  How much further along the line than the diamond is the circle? _____

d  Draw a triangle that is $\frac{1}{10}$ past the halfway position on the line.

# Independent practice

[Fraction wall from 1 whole down to 1/12]

**1** The fraction wall shows that $\frac{2}{4}$ is equivalent to $\frac{1}{2}$.

   **a** What other fractions on the fraction wall are equivalent to $\frac{1}{2}$?

   _____

   **b** Write a fraction that is equivalent to $\frac{1}{2}$ but is **not** on the fraction wall. $\frac{\square}{\square}$

**2** Find a fraction that is equivalent to:

   **a** $\frac{2}{10} = \frac{\square}{\square}$   **b** $\frac{1}{6} = \frac{\square}{\square}$   **c** $\frac{3}{12} = \frac{\square}{\square}$

   **d** $\frac{5}{6} = \frac{\square}{\square}$   **e** $\frac{4}{5} = \frac{\square}{\square}$   **f** $\frac{3}{9} = \frac{\square}{\square}$

   *Equivalent fractions are the same size.*

**3** Fill in the blanks.

   **a** $\frac{4}{6} = \frac{\square}{3}$   **b** $\frac{\square}{10} = \frac{4}{5}$   **c** $\frac{6}{\square} = \frac{3}{4}$

   **d** $\frac{6}{\square} = \frac{2}{3}$   **e** $\frac{2}{3} = \frac{\square}{12}$   **f** $\frac{3}{\square} = \frac{9}{12}$

We can show equivalent fractions in a diagram.

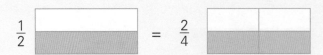

**4** Shade to show these fractions.

a Shade to show that $\frac{6}{8}$ is equivalent to $\frac{3}{4}$. Write the number sentence.

b Shade to show a fraction that is equivalent to $\frac{4}{5}$. Write the number sentence.

_____

**5** Divide and shade the shapes to show that:

a $\frac{3}{6}$ is equivalent to $\frac{1}{2}$.

b $\frac{6}{8}$ is equivalent to $\frac{3}{4}$.

**6**

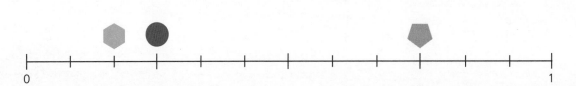

a The circle is $\frac{1}{4}$ of the way along the number line. Write another fraction that describes its position. _____

b Apart from $\frac{9}{12}$, what other fraction describes the position of the pentagon?

_____

c Write two equivalent fractions to describe the position of the hexagon. _____

d Draw a star $\frac{2}{3}$ of the way along the line.

# Extended practice

**1** If you look at an equivalent fraction such as $\frac{2}{4} = \frac{1}{2}$, you can see that there is a connection between the numerator and denominator.

What is the connection between the **numerator** and the **denominator** in each of these pairs of fractions?

a ÷4  $\frac{4}{8} = \frac{1}{2}$

b  $\frac{3}{9} = \frac{1}{3}$

c  $\frac{4}{10} = \frac{2}{5}$

d  $\frac{9}{12} = \frac{3}{4}$

e  $\frac{5}{10} = \frac{1}{2}$

f  $\frac{8}{12} = \frac{2}{3}$

g  $\frac{2}{5} = \frac{4}{10}$

h  $\frac{3}{4} = \frac{6}{8}$

i  $\frac{1}{3} = \frac{4}{12}$

j  $\frac{4}{5} = \frac{8}{10}$

k  $\frac{1}{2} = \frac{6}{12}$

l  $\frac{2}{3} = \frac{8}{12}$

**2** Choose a method to find a fraction that is equivalent to each of these:

a  $\frac{4}{6}$

b  $\frac{15}{20}$

c  $\frac{9}{18}$

d  $\frac{8}{20}$

e  $\frac{2}{14}$

f  $\frac{8}{10}$

g  $\frac{25}{100}$

**3** Reduce these fractions to their lowest equivalent form.

a  $\frac{8}{16}$

b  $\frac{16}{20}$

c  $\frac{8}{24}$

d  $\frac{9}{27}$

e  $\frac{6}{36}$

f  $\frac{80}{100}$

# UNIT 2: TOPIC 2
## Adding and subtracting fractions

Adding and subtracting *like* fractions (such as $\frac{3}{4} - \frac{1}{4}$) is as easy as working out 3 jelly beans – 1 jelly bean.

With *unlike* fractions, use your knowledge of *equivalent* fractions to add or subtract.

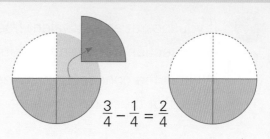

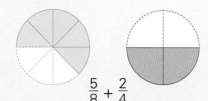

## Guided practice

**1 a**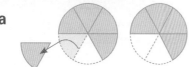

$\frac{5}{6} - \frac{1}{6} = \frac{\Box}{6}$

**b** $\frac{3}{7} + \frac{5}{7} = \frac{\Box}{7} = \underline{\phantom{0}}\frac{\Box}{7}$

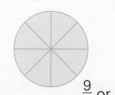 

*Remember that fractions need to be like fractions for addition and subtraction.*

**c**

$\frac{3}{4} + \frac{3}{4} = \frac{\Box}{4} = \underline{\phantom{0}}\frac{\Box}{\Box}$

**d**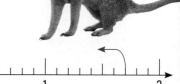

$1\frac{7}{10} - \frac{9}{10} = \frac{\Box}{10}$

**2 a**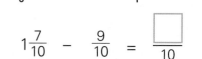

$\frac{3}{8} + \frac{1}{4} = \frac{3}{8} + \frac{\Box}{8} = \frac{\Box}{8}$

**b**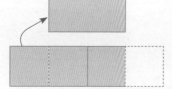

$\frac{3}{4} - \frac{1}{2} = \frac{3}{4} - \frac{\Box}{4} = \frac{\Box}{4}$

## Independent practice

**1**
a) $\frac{3}{8} + \frac{2}{8} =$ _____
b) $\frac{5}{10} + \frac{3}{10} =$ _____
c) $\frac{2}{5} + \frac{2}{5} =$ _____

d) $\frac{3}{7} + \frac{2}{7} =$ _____
e) $\frac{7}{12} + \frac{3}{12} =$ _____
f) $\frac{2}{9} + \frac{5}{9} =$ _____

g) $\frac{1}{6} + \frac{4}{6} =$ _____
h) $\frac{5}{10} + \frac{5}{10} =$ _____
i) $\frac{3}{8} + \frac{4}{8} =$ _____

**2**
a) $\frac{7}{8} - \frac{3}{8} =$ _____
b) $\frac{8}{9} - \frac{2}{9} =$ _____
c) $\frac{11}{12} - \frac{5}{12} =$ _____

d) $\frac{3}{4} - \frac{2}{4} =$ _____
e) $\frac{5}{7} - \frac{3}{7} =$ _____
f) $\frac{9}{10} - \frac{3}{10} =$ _____

g) $\frac{8}{9} - \frac{4}{9} =$ _____
h) $\frac{5}{6} - \frac{2}{6} =$ _____

**3**
a) $\frac{2}{8} + \frac{1}{4} =$ _____
b) $\frac{3}{10} + \frac{2}{5} =$ _____

c) $\frac{1}{3} + \frac{1}{6} =$ _____
d) $\frac{3}{4} + \frac{1}{8} =$ _____

e) $\frac{7}{10} + \frac{1}{5} =$ _____
f) $\frac{2}{9} + \frac{1}{3} =$ _____

*Remember to find a common denominator when adding or subtracting unlike fractions.*

**4**
a) $\frac{3}{8} - \frac{1}{4} =$ _____
b) $\frac{8}{10} - \frac{1}{2} =$ _____

c) $\frac{7}{12} - \frac{1}{4} =$ _____
d) $\frac{8}{9} - \frac{1}{3} =$ _____

e) $\frac{3}{4} - \frac{1}{2} =$ _____
f) $\frac{9}{12} - \frac{1}{4} =$ _____

**5** Shade the diagram to solve the addition problem. Write a number sentence that matches the problem.

 +  =  +  =

_____

**6** Use improper fractions and mixed numbers if the answer is greater than one whole. For example, $\frac{3}{4} + \frac{2}{4} = \frac{5}{4} = 1\frac{1}{4}$

*You could use pictures or number lines to help with these.*

a   $\frac{7}{8} + \frac{5}{8} =$ _____

b   $\frac{5}{9} + \frac{7}{9} =$ _____

c   $\frac{8}{12} + \frac{8}{12} =$ _____

d   $\frac{3}{4} + \frac{3}{4} =$ _____

e   $\frac{7}{10} + \frac{9}{10} =$ _____

f   $\frac{5}{6} + \frac{4}{6} =$ _____

g   $\frac{5}{8} + \frac{3}{8} =$ _____

h   $\frac{2}{3} + \frac{2}{3} =$ _____

**7** Solve these.

a   $1\frac{7}{8} - \frac{3}{8} =$ _____

b   $1\frac{5}{9} - \frac{4}{9} =$ _____

c   $1\frac{7}{10} - \frac{5}{10} =$ _____

d   $2\frac{3}{4} - \frac{1}{4} =$ _____

e   $3\frac{5}{8} - \frac{7}{8} =$ _____

f   $4\frac{8}{10} - \frac{9}{10} =$ _____

g   $2\frac{1}{9} - \frac{8}{9} =$ _____

h   $3\frac{3}{8} - \frac{5}{8} =$ _____

**8** Solve these.

a   $\frac{7}{8} + \frac{1}{4} =$ _____

b   $\frac{9}{10} + \frac{1}{5} =$ _____

c   $1\frac{2}{3} + \frac{5}{6} =$ _____

d   $2\frac{3}{4} + \frac{5}{8} =$ _____

e   $1\frac{7}{10} + \frac{4}{5} =$ _____

f   $3\frac{7}{9} + \frac{2}{3} =$ _____

**9** Solve these.

a   $1\frac{5}{8} - \frac{3}{4} =$ _____

b   $2\frac{3}{10} - \frac{1}{2} =$ _____

c   $1\frac{5}{12} - \frac{1}{2} =$ _____

d   $1\frac{2}{3} - \frac{5}{6} =$ _____

e   $2\frac{3}{4} - 1\frac{1}{2} =$ _____

f   $1\frac{1}{12} - \frac{2}{3} =$ _____

# Extended practice

**1** From the way this cake is arranged, it is obvious that the slices will add together to make a whole cake:

$\frac{5}{18}$  $\frac{2}{9}$  $\frac{1}{6}$  $\frac{1}{3}$

Prove that these pieces can also be added together to make a whole cake. (Hint: Look at the size of each fraction.)

Working-out space

**2** Write each answer in its **simplest form**.

a  $\frac{9}{10} + \frac{3}{5} =$ _____

b  $1\frac{5}{6} + \frac{7}{12} =$ _____

c  $3\frac{1}{4} - 1\frac{5}{8} =$ _____

d  $1\frac{13}{100} - \frac{1}{10} =$ _____

e  $2\frac{5}{12} + 3\frac{4}{12} =$ _____

f  $3\frac{2}{3} - 2\frac{1}{6} =$ _____

g  $2\frac{1}{3} + 1\frac{1}{4} =$ _____

**3** At a party, there are parts of four cakes left over. One cake was split into quarters, but the others were split into different fractions. The total amount left is one and one-sixth. What fraction of each cake might be left?

Working-out space

# UNIT 2: TOPIC 3
## Decimal fractions

**Common decimal fractions**

The most common decimal fractions are tenths, hundredths and thousandths.

If this is one whole ... 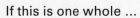 ... this is one-tenth

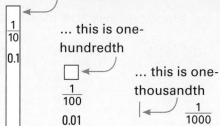

## Guided practice

**1**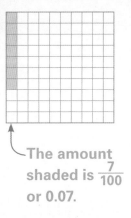

The amount shaded is $\frac{7}{100}$ or 0.07.

a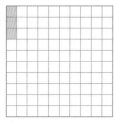

The amount shaded is _____

b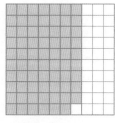

The amount shaded is _____

c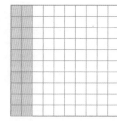

The amount shaded is _____

**2**

1000 jelly beans

If you had 500 of the jelly beans, you could call them a half, $\frac{500}{1000}$ or 0.5.

How many of the jelly beans do these fractions mean?

a  0.002  _____

b  0.008  _____

c  0.125  _____

d  $\frac{200}{1000}$  _____

e  $\frac{75}{1000}$  _____

f  0.009  _____

g  0.099  _____

h  0.999  _____

i  0.001  _____

j  0.01  _____

k  0.1  _____

l  0.25  _____

## Independent practice

**1** Shade:

a  0.05

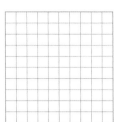

b  0.35

c  $\frac{33}{100}$

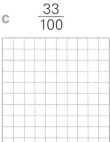

d  0.9

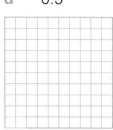

**2** Write **True** or **False** next to each of these:

a  0.5 > 0.05  _____

b  $\frac{7}{1000}$ < 0.007  _____

c  $\frac{17}{100}$ = 0.17  _____

d  0.009 > 0.01  _____

e  $\frac{175}{1000}$ = 0.175  _____

f  $\frac{1}{4}$ > 0.025  _____

g  0.04 = $\frac{4}{1000}$  _____

h  1.001 > 0.99  _____

i  3.25 = $3\frac{1}{4}$  _____

j  5.052 > 5.502  _____

k  2.430 > 2.43  _____

l  9.999 < 10  _____

**3**

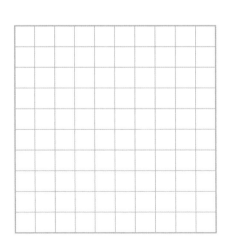

a  Colour 0.15 red.

b  Colour 0.05 yellow.

c  Colour 0.45 blue.

d  Colour one-tenth green.

e  Write the unshaded amount as a fraction and as a decimal.

_____

**4** Order these from **smallest** to **largest**:

0.45   0.145   0.415   0.451   0.045

_____   _____   _____   _____   _____

**5** Complete this table. Write the missing fractions and decimals.

| | Fraction | Decimal |
|---|---|---|
| | $\frac{1}{2}$ | 0.5 |
| a | $\frac{3}{4}$ | |
| b | | 0.1 |
| c | | 0.30 |
| d | $\frac{9}{100}$ | |
| e | | 0.405 |
| f | $\frac{250}{1000}$ | |
| g | $\frac{99}{1000}$ | |
| h | | 0.01 |

*Remember, the first decimal column is for tenths, the second is for hundredths and the third is for thousandths.*

**6** Change these improper fractions to mixed numbers and then to decimals.

| | Improper fraction | Mixed number | Decimal |
|---|---|---|---|
| | $\frac{5}{4}$ | $1\frac{1}{4}$ | 1.25 |
| a | $\frac{7}{4}$ | | |
| b | $\frac{13}{10}$ | | |
| c | $\frac{125}{100}$ | | |
| d | $\frac{450}{100}$ | | |
| e | $\frac{275}{100}$ | | |
| f | $\frac{1250}{1000}$ | | |

# Extended practice

**1** Some common fractions look just as simple when they are expressed as a decimal. Convert these fractions to decimals, and the decimals to fractions.

a  $\frac{1}{10} = $ _____

b  $\frac{1}{4} = $ _____

c  $\frac{7}{10} = $ _____

d  0.01 = _____

e  0.75 = _____

f  0.001 = _____

**2** To change a fraction to a decimal, divide the numerator by the denominator.

$$2 \overline{) 1.^{1}0 }\phantom{xx}^{0.\phantom{x}5}$$

Change these to decimals.

a  $\frac{1}{5}$

b  $\frac{1}{8}$

c  $\frac{3}{4}$

d  $\frac{3}{8}$

e  $\frac{4}{5}$

f  $\frac{7}{8}$

**3** Another common fraction is $\frac{1}{3}$, but it doesn't look simple when it is expressed as a decimal. Find the decimal equivalent of $\frac{1}{3}$ by writing an algorithm or using a calculator.

**4** Decimals in which a number is repeated over and over are called *recurring decimals*. To show the recurring number in a decimal, you can place a dot (like a full stop) over the top of the number that recurs. Find the decimal equivalent of $\frac{1}{6}$, then place a dot over the recurring number.

**5** Some fractions convert to a very long decimal. Find the decimal equivalent of $\frac{1}{7}$, then round it to an appropriate number of places.

# UNIT 2: TOPIC 4
## Addition and subtraction of decimals

You can add or subtract decimals just like you do with whole numbers:

|   | 3 | 1 | 4 |
|---|---|---|---|
| + | 1 | 7 | 3 |
|   | 4 | 8 | 7 |

|   | 3 • | 1 | 4 |
|---|---|---|---|
| + | 1 • | 7 | 3 |
|   | 4 • | 8 | 7 |

But if there is a different number of columns, it is important to line up the numbers according to their place value:

|   | 2 | 3 | 1 | 7 |
|---|---|---|---|---|
| + |   | 5 | 9 | 7 |
|   | 2 | 9 | 1 | 4 |

|   | 2 | 3 • | 1 | 7 |
|---|---|---|---|---|
| + | 5 | 9 • | 7 |   |
|   | 8 | 2 • | 8 | 7 |

*The decimal point doesn't make much difference to the way you work, but it makes a BIG difference to the answer.*

## Guided practice

**1** Find the answers.

a
|   | 2 | 5 | 3 | 7 |
|---|---|---|---|---|
| + | 1 | 6 | 2 | 9 |
|   |   |   |   |   |

b
|   | 2 | 5 • | 3 | 7 |
|---|---|---|---|---|
| + | 1 | 6 • | 2 | 9 |
|   |   | • |   |   |

**2** Use place value to line up the numbers and calculate the answers.

a   32.8 + 12.4

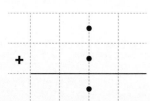

b   2.47 + 1.9

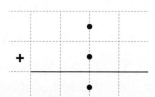

c   24.74 + 4.38

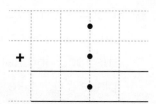

d   75.9 − 23.6

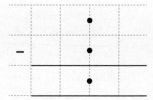

e   4.45 − 2.7

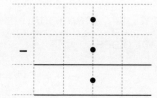

f   36.25 − 9.28

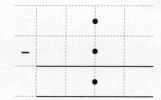

# Independent practice

**1** Calculate the answers.

a
```
    2 . 5 4
+   3 . 4 8
_____
      .
```

b
```
    4 . 3 9
+   4 . 9 7
_____
      .
```

c
```
  3 5 . 1 8 7
+ 2 8 . 7 4 9
_____
       .
```

d
```
    8 2 . 5
-   3 2 . 4
_____
      .
```

e
```
    4 . 2 8
-   2 . 7 3
_____
      .
```

f
```
  4 3 . 0 5 6
- 3 5 . 4 6 3
_____
       .
```

g   7.45 − 5.24

h   42.7 − 32.8

i   46.7 − 29.285

**2**  a   Add $23.79 and $147.35.

b   Subtract $119.95 from $200.

**3**  a   Find the total of 2.54 m, 17.7 m and 34.67 m.

b   By how much is 3.463 kg less than $5\frac{3}{4}$ kg?

**4** Sam can run two 50-metre laps of the school athletic track in less than 18 seconds. What is the most likely time for each lap?

_____

a   82.53 seconds

b   9.253 seconds

c   92.53 seconds

d   8.253 seconds

**5** Bill is building a fence that is 73.17 m long. He has already finished thirty-nine and a quarter metres of it. How much more does he need to build?

Working-out space

**6** Find the total mass of a parcel that has four items that weigh:
4.45 kg, 3.325 kg, $1\frac{1}{2}$ kg, 725 g

Working-out space

**7** A roll of cloth is 14.36 m long. How much is left after $5\frac{3}{4}$ metres have been cut from the roll?

Working-out space

# Extended practice

**1** The answer to this equation is 9.18. Try to find at least two ways of filling the gaps to complete the equation.

a   0.☐ + 4.☐2 + ☐.36 = 9.18        Working-out space

b   0.☐ + 4.☐2 + ☐.36 = 9.18        Working-out space

---

**2** Did you know that your skin weighs almost as much as your bones? This table lists the mass of the eight largest organs in an adult who weighs 68 kg.

a   Rewrite the table, listing the organs from **heaviest** to **lightest**.

| Organ | Mass |
|---|---|
| heart | 0.315 kg |
| lungs | 1.09 kg |
| skin | 10.886 kg |
| pancreas | 0.098 kg |
| brain | 1.408 kg |
| spleen | 0.17 kg |
| liver | 1.56 kg |
| kidneys | 0.29 kg |

| Organ | Mass |
|---|---|
|  |  |
|  |  |
|  |  |
|  |  |
|  |  |
|  |  |
|  |  |
|  |  |

b   Find the total mass of the heart and lungs. _____

c   How much heavier is the skin than the brain? _____

d   The mass of which organ is closest to the mass of the kidneys? _____

e   The right lung is 0.07 kg heavier than the left lung (to make space for the heart). What might the two masses be? _____

f   What is the difference between the mass of the lungs and the mass of the pancreas? _____

g   The mass of an adult male gorilla is about 240 kg, but his brain weighs only 0.465 kg. How much heavier is a human brain than a gorilla's brain? _____

# UNIT 2: TOPIC 5
## Multiplication and division of decimals

You can multiply decimals in the same way that you multiply whole numbers:

**Multiplying a whole number by 4**

Extended:
```
    2 4
  ×   4
  ─────
    1 6
  + 8 0
  ─────
    9 6
```

Short:
```
    2 4
  ×   4
  ─────
    9 6
```

**Multiplying a decimal by 4**

Extended:
```
    2•4
  ×   4
  ─────
    1•6
  + 8•0
  ─────
    9•6
```

Short:
```
    2•4
  ×   4
  ─────
    9•6
```

## Guided practice

**1** a
```
    1 3 2
  ×     3
  ───────
```
b
```
    1 3•2
  ×     3
  ───────
       •
```

> You could do these as extended multiplications on a separate piece of paper.

**2** a
```
    2 1 3 5
  ×       4
  ─────────
```
b
```
    2 1•3 5
  ×       4
  ─────────
         •
```

**3** a
```
    4 2 6
  ×     7
  ───────
```
b
```
    4 2•6
  ×     7
  ───────
       •
```

**4** a
```
    3 0 7 3
  ×       6
  ─────────
```
b
```
    3 0•7 3
  ×       6
  ─────────
         •
```

**Dividing a whole number by 3 and a decimal by 3:**

```
      8 3              8.3
  3 )2 4 9    →    3 )2 4.9
```

**5** a  3 )5 1 6    b  3 )5 1.6
**6** a  5 )8 5 5    b  5 )8 5.5

**7** a  7 )5 7 4    b  7 )5 7.4
**8** a  4 )8 1 6    b  4 )8 1.6

## Independent practice

*Remember to put the decimal point in the correct place!*

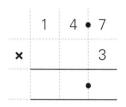

**1**

a) 14.7 × 3

b) 21.4 × 4

c) 31.5 × 2

d) 18.3 × 5

e) 4.32 × 2

f) 2.59 × 4

g) 1.35 × 6

h) 2.57 × 3

i) 12.95 × 4

j) 432.1 × 2

k) 2.575 × 3

l) 13.59 × 7

**2**

a) 3)15.9

b) 6)72.6

c) 4)49.6

d) 5)97.5

e) 5)5.25

f) 4)9.48

g) 2)6.38

h) 7)84.7

i) 4)57.52

j) 3)37.41

k) 8)20.56

l) 9)4.743

**3** Multiply:

a   43.6 by 4

b   54.6 by 6

c   7.39 by 5

d   42.67 by 3

e   46.32 by 7

f   7.456 by 4

g   90.25 by 8

h   62.05 by 9

i   8.035 by 5

**4** Often items are sold at a price that is impossible to pay with an exact number of coins. If you saw pens at $1.99 each, how much would you actually pay for:

a   1 pen? _____

b   2 pens? _____

c   4 pens? _____

d   10 pens? _____

Working-out space

**5** Find the cost of **each item** in these packs. Round each answer to an appropriate amount of money.

a   Two toys for $1.99   _____

b   Five party hats for $7.99   _____

c   Three prizes for $8.99   _____

d   Four drinks for $4.99   _____

# Extended practice

**1** Pete's Pizza Place gets an order for eight ordinary pizzas at $8.95 each and one pizza supreme at $12.95. What is the total cost?

**2** A 5.4 m length of wood is cut into nine equal pieces. How long is each piece?

**3** Eight people share a prize of $500. How much does each person receive?

**4** Here is a list of items bought for a party for a group of six students:

| Item | Unit cost | Number required | Cost |
| --- | --- | --- | --- |
| Soft drink (1.25 L) | $2.25 | half a bottle for each student | |
| Juice (300 mL) | $0.84 | one for each student | |
| Potato crisps (50 g) | $1.35 | two for each student | |
| Chocolate (150 g) | $4.93 | two packets | |
| Melon | $3.84 | one between the group | |
| Pies (4 in a pack) | $8.04 | one pie for each student | |

**a** Fill in the cost column for each row. Consider whether you will need to round the amounts of money.

**b** What is the cost of all the items for the group? _____

**c** How much per person does it cost for the melon? _____

**d** What is the total cost for **each** of the six students? _____

**e** There are four groups of six in the class. What is the cost for the whole class?

Working-out space

# UNIT 2: TOPIC 6
## Decimals and powers of 10

Multiplying a decimal by 10 is almost the same as multiplying a whole number by 10: everything moves one place bigger.

The difference is the **zero**. You have to decide whether it is needed.

34.0 has the same value as 34, so you can write 3.4 × 10 = 34

## Guided practice

1. a  45 × 10 = _____
   b  4.5 × 10 = _____

2. a  74 × 10 = _____
   b  7.4 × 10 = _____

3. a  375 × 10 = _____
   b  37.5 = _____

4. a  629 × 10 = _____
   b  62.9 × 10 = _____

*We don't usually put a zero unless it is necessary.*

Dividing by 10 moves every digit the opposite way:

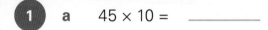   

5. a  350 ÷ 10 = _____
   b  35 ÷ 10 = _____

6. a  740 ÷ 10 = _____
   b  74 ÷ 10 = _____

7. a  870 ÷ 10 = _____
   b  87 ÷ 10 = _____

8. a  930 ÷ 10 = _____
   b  93 ÷ 10 = _____

9. a  32.6 × 10 _____
   b  2.35 × 10 _____
   c  7.892 × 10 _____
   d  65.2 × 10 _____

10. a  23.5 ÷ 10 _____
    b  42.75 ÷ 10 _____
    c  3.5 ÷ 10 _____
    d  0.2 ÷ 10 _____

# Independent practice

Multiplying by 100 moves each digit **two** places larger:

space-filler

Solve these multiplication problems.

**1**  a   3.5 × 10 = _____

   b   3.5 × 100 = _____

**2**  a   6.7 × 10 = _____

   b   6.7 × 100 = _____

**3**  a   5.38 × 10 = _____

   b   5.38 × 100 = _____

**4**  a   4.09 × 10 = _____

   b   4.09 × 100 = _____

Dividing by 100 moves each digit **two** places smaller:

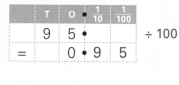

Solve these division problems.

**5**  a   4.5 ÷ 10 = _____

   b   4.5 ÷ 100 = _____

**6**  a   7.9 ÷ 10 = _____

   b   7.9 ÷ 100 = _____

**7**  a   54.5 ÷ 10 = _____

   b   54.5 ÷ 100 = _____

**8**  a   62.7 ÷ 10 = _____

   b   62.7 ÷ 100 = _____

Solve these multiplication problems.

**9**  a   2.45 × 100 = _____     b   17.37 × 100 = _____

Solve these division problems.

**10**  a   3416.1 ÷ 100 = _____     b   0.1 ÷ 100 = _____

OXFORD UNIVERSITY PRESS

Multiplying or dividing by 1000 moves the digits over **three** places.

**Multiplying by 1000**

| | Th | H | T | O | . | 1/10 |
|---|---|---|---|---|---|---|
| | | | | 3 | . | 7 |
| = | 3 | 7 | 0 | 0 | | |

space-fillers

× 1000

**Dividing by 1000**

| | H | T | O | . | 1/10 | 1/100 | 1/1000 |
|---|---|---|---|---|---|---|---|
| | 1 | 4 | 2 | . | | | |
| = | | | 0 | . | 1 | 4 | 2 |

÷ 1000

**11** Multiply by 1000.

| | | Tth | Th | H | T | O | . | 1/10 |
|---|---|---|---|---|---|---|---|---|
| a | 1.3 | | 1 | | | | . | |
| b | 2.6 | | | | | | . | |
| c | 3.57 | | | | | | . | |
| d | 1.27 | | | | | | . | |
| e | 15.47 | | | | | | . | |
| f | 72.95 | | | | | | . | |
| g | 96.3 | | | | | | . | |
| h | 25.4 | | | | | | . | |

**12** Divide by 1000.

| | | H | T | O | . | 1/10 | 1/100 | 1/1000 |
|---|---|---|---|---|---|---|---|---|
| a | 432 | | | 0 | . | | | |
| b | 529 | | | | . | | | |
| c | 841 | | | | . | | | |
| d | 697 | | | | . | | | |
| e | 1485 | | | | . | | | |
| f | 3028 | | | | . | | | |
| g | 10 436 | | | | . | | | |
| h | 99 999 | | | | . | | | |

**13** Complete the tables.

| | | × 10 | × 100 | × 1000 |
|---|---|---|---|---|
| a | 1.7 | | | |
| b | 22.95 | | | |
| c | 3.02 | | | |
| d | 4.42 | | | |
| e | 5.793 | | | |
| f | 21.578 | | | |
| g | 33.008 | | | |
| h | 29.005 | | | |

**14**

| | | ÷ 10 | ÷ 100 | ÷ 1000 |
|---|---|---|---|---|
| a | 74 | | | |
| b | 7 | | | |
| c | 18 | | | |
| d | 325 | | | |
| e | 2967 | | | |
| f | 3682 | | | |
| g | 14 562 | | | |
| h | 75 208 | | | |

# Extended practice

**1** One of these processes will give the correct answer to 2.25 × 0.4. Which one is it?

☐ 225 × 4 × 100    ☐ 2.25 × 4 ÷ 100    ☐ 225 × 4 ÷ 1000    ☐ 2.25 × 4 ÷ 1000

**2** Use what you found out in question 1 to find the answer to these problems.

a  3.12 × 0.3  _____    b  31.2 × 0.3  _____

c  20.3 × 0.03  _____    d  40.02 × 0.2  _____

**3** How many jumps of 0.2 on a number line would take you from 0 to 500? _____

**4** A fast food store has a 150-litre barrel of juice. How many cups can be filled if the cup sizes are:

a  0.25 L?  _____    c  0.15 L?  _____

b  0.2 L?  _____    d  600 mL?  _____

**5** A shop paid $132 948 for 1000 watches.

a  What was the average price of each watch?  _____

b  One-tenth of the total price was for insurance. What was the insurance charge?  _____

c  One watch was worth one-hundredth of the total price. How much was that watch?  _____

Working-out space

# UNIT 2: TOPIC 7
## Percentage, fractions and decimals

The symbol % stands for *per cent*. It means *out of a hundred*. So 1% means 1 out of 100. It can be written as:

- a fraction: $\frac{1}{100}$
- a decimal: 0.01
- a percentage: 1%

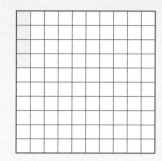

The amount shaded is:

$\frac{4}{100}$ (fraction)

0.04 (decimal)

4% (percentage)

## Guided practice

**1** Write a fraction, decimal and percentage for each shaded part.

a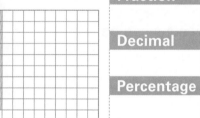
Fraction
Decimal
Percentage

b
Fraction
Decimal
Percentage

c
Fraction
Decimal
Percentage

d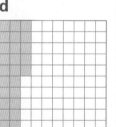
Fraction
Decimal
Percentage

e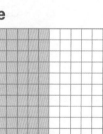
Fraction
Decimal
Percentage

f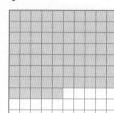
Fraction
Decimal
Percentage

**2** Shade the grid and fill the gaps.

a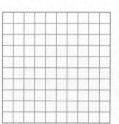
Fraction $\frac{2}{100}$
Decimal
Percentage

b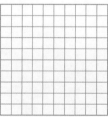
Fraction
Decimal 0.2
Percentage

c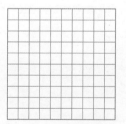
Fraction
Decimal
Percentage 35%

d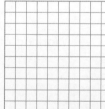
Fraction $\frac{7}{10}$
Decimal
Percentage

## Independent practice

*Remember that 1% has the same value as 0.01 and $\frac{1}{100}$.*

**1** Complete this table.

|   | Fraction | Decimal | Percentage |
|---|---|---|---|
| a | $\frac{15}{100}$ |   |   |
| b |   | 0.22 |   |
| c |   |   | 60% |
| d |   | 0.09 |   |
| e | $\frac{9}{10}$ |   |   |
| f |   |   | 53% |
| g |   | 0.5 |   |
| h | $\frac{1}{4}$ |   |   |
| i |   | 0.04 |   |
| j |   |   | 75% |
| k | $\frac{1}{5}$ |   |   |

**2** Write **True** or **False**.

a   30% = 0.3   _____

b   0.04 < 40%   _____

c   0.12 > $\frac{12}{100}$   _____

d   25% = $\frac{1}{4}$   _____

e   $\frac{3}{4}$ < 75%   _____

f   0.9 = 9%   _____

g   $\frac{2}{10}$ > 20%   _____

h   95% = 0.95   _____

i   100% = 1   _____

**3** Order these from **smallest** to **largest**.

a   0.3   20%   $\frac{1}{4}$   _____

b   0.07   69%   $\frac{6}{10}$   _____

c   17%   0.2   $\frac{2}{100}$   _____

d   $\frac{1}{4}$   4%   0.14   _____

e   10%   $\frac{1}{5}$   0.5   _____

f   39%   0.395   $\frac{3}{10}$   _____

**4** Find the matching fractions, decimals and percentages. Choose colours to lightly shade each matching set of three.

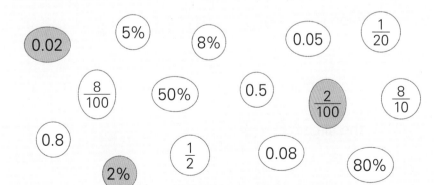

**5** Fill in the blanks on this number line.

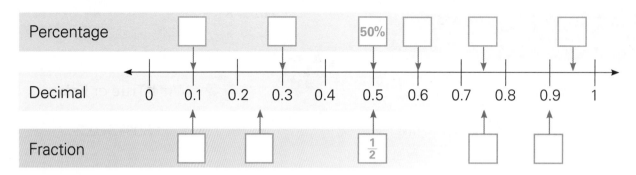

**6** Circle the percentage that matches the position of each shape.

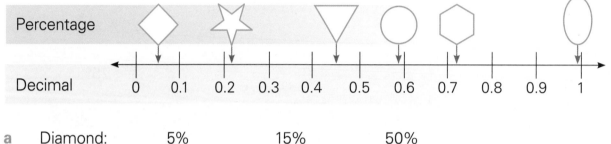

a Diamond: 5% 15% 50%
b Star: 15% 20% 22%
c Triangle: 40% 44% 49%
d Circle: 59% 55% 51%
e Hexagon: 70% 72% 75%
f Oval: 99% 100% 90%

**7** Draw a smiley face and an arrow 85% of the way along the number line in question 6.

**8** The square is 10% of the way along this number line. To the nearest 10%, what is the position of:

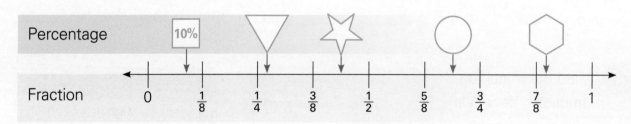

a the triangle? _____ b the star? _____

c the circle? _____ d the hexagon? _____

# Extended practice

Read the information about Australia. Then write your answers in complete sentences and you will have six facts about Australia. You might need to use spare paper for your working out.

**1** In Australia, there are around 28 million cattle. That is about $\frac{1}{50}$ of all the cattle in the world. What percentage of the world's cattle is in Australia? _____

_____

**2** There are 378 mammal species in Australia. 80% of them are found nowhere else in the world. Change 80% to a fraction and write it in its simplest form. _____

*Australia sounds like an interesting place. I might move there!*

**3** Around 25% of the people in Australia live in Victoria. Australia's population reached 22 million people in 2009. What was the approximate population of Victoria in 2009? _____

**4** Australia has 79 million sheep. This is $\frac{3}{20}$ of the number of sheep in the world's Top 10 sheep countries. What percentage of the sheep in the Top 10 countries are in Australia?

_____

**5** Some people think Australia is mainly desert. In fact, the Great Sandy Desert only covers about 5% of Australia. Write the fraction (in its simplest form) of Australia that is covered by the Great Sandy Desert.

_____

**6** Australia has 749 out of 5594 of the world's threatened animal species. Circle one answer. The percentage of the world's threatened animal species that are in Australia is about:

- 1%.
- 3%.
- 8%.
- 13%.

# UNIT 3: TOPIC 1
# Ratios

Ratios are used to compare numbers or quantities to each other.

In the example below, there are 6 smiley faces and 4 sad faces.

The ratio of smiley faces to sad faces is 6 to 4. This is written as 6:4.

## Guided practice

**1** Write the ratio of smiley faces to sad faces.

| | | Ratio of smiley faces to sad faces |
|---|---|---|
| | 😊😊😊☹️☹️😊😊☹️☹️ | 6:4 |
| a | | |
| b | | |
| c | | |

**2** In the first example, the ratio of 6:4 can be simplified. The simplest form of the ratio is 3:2 because there are 3 smiley faces for every 2 sad faces.

Write the ratio of smiley faces to sad faces below in its simplest form.

| | | Ratio in its simplest form |
|---|---|---|
| | | 3:2 |
| a | | |
| b | | |
| c | | |
| d | | |
| e | | |
| f | | |
| g | | |

*Finding the simplest ratio is like finding the lowest equivalent form in fractions.*

# Independent practice

**1** In a pack of pink and purple jellybeans, the ratio of pink to purple is 2:3. This means that there are 2 pink jellybeans for every 3 purple jellybeans.

If there are 8 pink jellybeans in total, we can use the ratio of 2:3 to work out the number of purple ones. Drawing the jellybeans can also help.

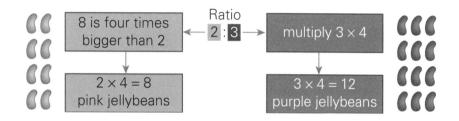

Use the ratio of 2:3 to work out the number of **purple** jellybeans in a jar if there are:

a  6 pink. _____     b  10 pink. _____     c  16 pink. _____

**2** Use the ratio of 2:3 to work out the number of **pink** jellybeans in a jar if there are:

a  6 purple. _____     b  10 purple. _____     c  16 purple. _____

**3** Ratios can compare more than two numbers. For every blue square in this pattern, there are 2 yellow squares and 3 green squares. The ratio is 1:2:3.

What is the ratio (in its simplest form) of blue to yellow to green squares below?

a  The ratio is _____.

b  The ratio is _____.

c  The ratio is _____.

**4** Colour the grid so that the ratio is 3 blue squares to 1 yellow square to 2 green squares (3:1:2).

**5** Look at this bead pattern.

Describe the bead pattern:

a in words. _____

b as a ratio. _____

**6** There are 24 beads in this necklace. Choose a ratio to make a pattern using the colours red, green and blue.

Describe the bead pattern:

a in words. _____

b as a ratio. _____

**7** To make 8 pancakes, Joe uses 120 g of flour, 250 mL of milk and 1 egg.
Using ratio, complete the table to help Joe work out the different quantities.

| Flour | Milk | Eggs | Number of pancakes |
|---|---|---|---|
| 120 g | 250 mL | 1 | 8 |
| 240 g | | | |
| | | 4 | |
| | 1.5 L | | |
| | | | 4 |

**8** Kate has 18 sheep, 48 goats, 6 horses and 12 ducks.

a In its simplest form, write the ratio of Kate's sheep to the goats, horses and ducks.

_____

b Zoe has the same types of animals as Kate, and in the same ratio, but she only has 4 ducks. How many of each of the other animals does Zoe have?

_____

# Extended practice

## Proportion

Proportion is different to ratio. It compares one number to the *whole* group.

**1** The ratio of strawberries to bananas is 1:3.

To find the proportion of strawberries, we look at the total number of fruit (8).

Next, we look at the number of fruit that are strawberries (2).

The fraction of the group that are strawberries is $\frac{2}{8}$, which can be simplified to $\frac{1}{4}$.

So the proportion of strawberries is $\frac{1}{4}$.

The proportion can also be written as a percentage (25%) or a decimal (0.25).

Write the proportion of bananas as:

a  a fraction. _____   b  a percentage. _____   c  a decimal. _____

**2** In a box of 20, the ratio of oranges to apples is 1:4. We can use ratio and proportion to work out the number of oranges and apples.

Add 1 orange and 4 apples: 1 + 4 = 5. This means there are 5 "portions".

Proportion of oranges: $\frac{1}{5}$    Proportion of apples: $\frac{4}{5}$

One-fifth of 20 is 4, so there are 4 oranges.

Four-fifths of 20 is 4 lots of 4, so there are 16 apples.

How many oranges and apples are in each box if the total number is:

a  10?           b  25?           c  50?           d  35?

Oranges: _____   Oranges: _____   Oranges: _____   Oranges: _____

Apples: _____    Apples: _____    Apples: _____    Apples: _____

**3** Use the information to work out the number of oranges and apples in each box.

a  Fruit box Contents: 45 pieces Ratio of oranges to apples = 3:2
Oranges: _____
Apples: _____

b  Fruit box Contents: 56 pieces Ratio of oranges to apples = 3:4
Oranges: _____
Apples: _____

c  Fruit box Contents: 32 pieces Ratio of oranges to apples = 1:3
Oranges: _____
Apples: _____

d  Fruit box Contents: 72 pieces Ratio of oranges to apples = 3:5
Oranges: _____
Apples: _____

# UNIT 4: TOPIC 1
## Geometric and number patterns

Patterns are all around us. There is a pattern in the way these craft sticks are placed.

We could describe the pattern like this:
*For every pentagon you use 5 sticks.*

## Guided practice

**1** Fill the gaps in this table.

| | Pattern | Rule | How many sticks are used |
|---|---|---|---|
| a | | For every pentagon you need ___ sticks | 4 × ☐ = ☐ |
| b | | For every diamond you need ___ sticks | ☐ × ☐ = ☐ |

**2** Complete this table and write a rule for the pattern.

| Position | 1 | 2 | 3 | 4 | 5 | 6 | 7 | 8 | 9 |
|---|---|---|---|---|---|---|---|---|---|
| Number | 10 | 9.5 | 9 | 8.5 | | | | | |

Rule: _____

**3** We can also use a flowchart to show a mathematics rule.
Use this flowchart to find whether these numbers are divisible by 4.

a  124 _____

b  516 _____

c  4442 _____

**4** What question should be written in the diamond in this flowchart?

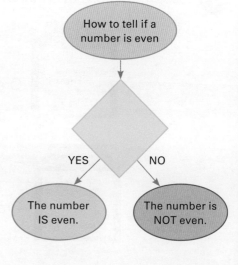

# Independent practice

**1** This pattern could **not** be described by the rule that you use four sticks for every square. Why not?

_____

To describe the pattern above, you need to look how it was made:

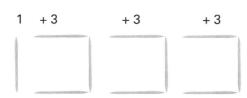

You start with one stick, then use 3 sticks for every square.

How many sticks were used for the three squares?   $1 + 3 \times 3 = 1 + 9 = 10$

**2** Complete this table.

| | Pattern | Rule | How many sticks are needed? |
|---|---|---|---|
| a | | Start with 1 stick, then use 3 sticks for every square. | $1 + 4 \times 3 = 1 + 12 = \square$ |
| b | | | $1 + \square \times 3 = 1 + \square = \square$ |
| c | | | |

**3** True or false? Both rules describe the pattern. _____

- Start with one stick and then use two sticks for every triangle.
- Use three sticks for the first triangle and then two for every other triangle.

*Try to make the rules easy to understand.*

**4** Write a simple rule for each pattern.

| | Pattern | Rule |
|---|---|---|
| a | | |
| b | | |
| c | | |
| d | | |

**5** Read the rule to complete each table.
Multiply the position number by 3 and then subtract 1.

| Position | 1 | 2 | 3 | 4 | 5 | 6 | 7 | 8 | 9 | 10 |
|---|---|---|---|---|---|---|---|---|---|---|
| Number | 2 | | | | | | | | | |

**6** Complete the table and write a rule for this number pattern.

| Position | 1 | 2 | 3 | 4 | 5 | 6 | 7 | 8 | 9 | 10 |
|---|---|---|---|---|---|---|---|---|---|---|
| Number | 1 | 4 | 9 | 16 | | | | | | |

**7** This flow chart shows the steps for doing a whole number division algorithm.

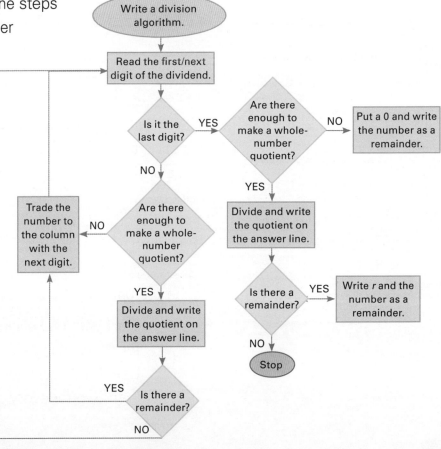

*The dividend is the number that is being divided.*

*The divisor is the number you are dividing by.*

*The quotient is the whole-number answer when a number has been divided.*

Follow the flow chart to complete these algorithms.

a  3)344  b  4)548  c  3)259

**8** To know if a number is divisible by 3, you add the digits together. If the answer is divisible by 3, then the whole number is divisible by 3. On a separate piece of paper, design a flow chart that shows the steps to find out whether a number is divisible by 3. Test it yourself before giving it to somebody to use.

# Extended practice

Imagine you are planning a sit-down party. How will the tables be arranged?
For all the following activities:
- only one person can sit along one side of a table
- use the abbreviation *n* for the number of people and *t* for the number of tables.

**1** A common shape for a table is rectangular. If the tables are separate, the formula for the number of people that can be seated is $n = t \times 4$.

Using this formula, how many can be seated at:

a  8 tables?    b  10 tables?    c  20 tables?    d  50 tables?

**2** At parties, the tables are often joined end to end.

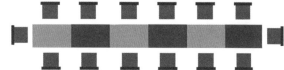

a  How many people could sit at 4 tables arranged like this? _____

b  Write a formula for the number of people who could sit at any number of tables arranged like this. _____

c  Rewrite the number of people who could sit at the numbers of tables in question 1.

a ☐    b ☐    c ☐    d ☐

**3** If the tables were like this, how many people could sit at:

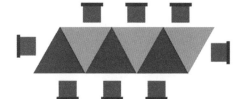

a  5 tables? _____    b  7 tables? _____

c  10 tables? _____    d  20 tables? _____

**4** a  Write a formula that would suit the seating arrangements in question 3.

b  Using the seating plan in question 3, how many tables would be needed to seat a class of 24 students? _____

# UNIT 4: TOPIC 2
## Order of operations

It doesn't matter which operation you do first in 2 + 5 − 3. The answer is still 4. But sometimes the order of operations *does* matter.

BODMAS is a way of knowing what to do first.

| | | | |
|---|---|---|---|
| 1st | B | Brackets | 2 × (3 − 1) = 4 |
| 2nd | O | Other operations | 4 × 3² = 4 × 9 = 36 <br> ½ of 10 + 4 = 5 + 4 = 9 |
| 3rd | D | Divide | 10 + 6 ÷ 2 = 10 + 3 = 13 |
| | M | Multiply | 2 × 3 + 2 = 6 + 2 = 8 |
| 4th | A | Add | 4 + 2 × 3 = 4 + 6 = 10 |
| | S | Subtract | 5 × 4 − 3 = 20 − 3 = 17 |

What is three times five plus two?

It's 17!
3 × 5 is 15
15 + 2 = 17

No, it's 21!
5 + 2 is 7
3 × 7 = 21

## Guided practice

**1** What is the correct answer to the question in the speech bubble above? _____

**2**
a  3 + 2 × 2 = 3 + 4 = _____
b  (3 + 2) × 2 = _____
c  6 × 4 − 3 = _____
d  6 × (4 − 3) = _____
e  48 ÷ 8 − 2 = _____
f  48 ÷ (8 − 2) = _____
g  8 + 12 ÷ 2 = _____
h  (8 + 12) ÷ 2 = _____

**3**
a  ½ of 8 × 3 = 4 × 3 = _____
b  ½ of (8 × 3) = _____
c  ½ of 6 + 3 = _____
d  ½ of (6 + 3) = _____
e  4² + 5 = _____
f  5² + 4 = _____
g  3 × 2² = _____
h  (3 × 2)² = _____

**4**
a  3 × (10 − 5) = _____
b  ¼ of 20 × 2 = _____
c  5 + 6 ÷ 2 = _____
d  ½ of 24 ÷ 6 = _____
e  (7² + 1) × 2 = _____
f  3 × 12 ÷ 2 = _____
g  ½ of 10 × 2² = _____
h  5 + (10 − 5)² = _____

# Independent practice

$5 \times 4 = 15 + 5$

An equation is a number sentence in two parts. The two parts balance each other.

**1** Complete these equations.

a  $5 \times 2 = \underline{\phantom{XX}} + 8$

b  $\underline{\phantom{XX}} \times 5 = 30 - 5$

c  $24 \div 2 = 4 \times \underline{\phantom{XX}}$

d  $\underline{\phantom{XX}} + 7 = (4 + 5) \times 3$

e  $\frac{1}{2}$ of $6 + 5 = 24 \div \underline{\phantom{XX}}$

**2** You can use equations to make multiplication simpler.
Use equations to split the number you are multiplying.

| | Problem | | Split the problem to make it simpler | | Solve the problem | | Answer |
|---|---|---|---|---|---|---|---|
| | 27 × 3 | = | (20 × 3) + (7 × 3) | = | 60 + 21 | = | 81 |
| a | 23 × 4 | = | (20 × 4) + (3 × 4) | = | | = | |
| b | 19 × 7 | = | | = | | = | |
| c | 48 × 5 | = | | = | | = | |
| d | 37 × 6 | = | | = | | = | |
| e | 29 × 5 | = | | = | | = | |
| f | 43 × 7 | = | | = | | = | |
| g | 54 × 9 | = | | = | | = | |

**3** Use equations to change the order of operations.

| | Problem | | Change the order to make it simpler | | Solve the problem | | Answer |
|---|---|---|---|---|---|---|---|
| | 20 × 17 × 5 | = | 20 × 5 × 17 | = | 100 × 17 | = | 1700 |
| a | 20 × 13 × 5 | = | 20 × 5 × 13 | = | | = | |
| b | 25 × 14 × 4 | = | | = | | = | |
| c | 5 × 19 × 2 | = | | = | | = | |
| d | 25 × 7 × 4 | = | | = | | = | |
| e | 60 × 12 × 5 | = | | = | | = | |
| f | 5 × 18 × 2 | = | | = | | = | |
| g | 25 × 7 × 8 | = | | = | | = | |

OXFORD UNIVERSITY PRESS

You can use "opposites" to solve problems. To find the value of ◊ in the equation ◊ + 3 = 9, move the + 3 to the other side and do the opposite of plus. It becomes ◊ = 9 − 3, so ◊ = 6. You can check it by writing the equation: 6 = 9 − 3

**4** Rewrite each equation using "opposites" to find the value of ◊.

| | Problem | Use opposites | Find the value of ◊ | Check by writing the equation |
|---|---|---|---|---|
| e.g. | ◊ + 15 = 35 | ◊ = 35 − 15 | ◊ = 20 | 20 = 35 − 15 |
| a | ◊ × 6 = 54 | ◊ = 54 ÷ 6 | | |
| b | ◊ + 1.5 = 6 | | | |
| c | $\frac{1}{4}$ of ◊ = 10 | | | |
| d | ◊ × 10 = 45 | | | |
| e | ◊ ÷ 10 = 3.5 | | | |
| f | ◊ ÷ 4 = 1.5 | | | |
| g | ◊ × 100 = 725 | | | |

**5** Another strategy to find the value of ◊ is to put a number in its place to see if it balances the equation. The number is a "substitute" for ◊. For example, ◊ + $4^2$ = 18. Substitute 2 for ◊. Does 2 + $4^2$ = 18? Yes! So, ◊ = 2.

| | Problem | Possible substitutes for ◊ | | | | Check |
|---|---|---|---|---|---|---|
| e.g. | $◊^2$ × 3 = 75 | 4 | 5 | 6 | 7 | $5^2$ × 3 = 25 × 3 = 75 |
| a | ◊ × 3 + 5 = 32 | 8 | 9 | 10 | 11 | |
| b | 54 ÷ ◊ − 5 = 1 | 9 | 10 | 11 | 12 | |
| c | 2 × ◊ + 5 = 15 | 2 | 3 | 4 | 5 | |
| d | 15 ÷ ◊ − 1.5 = 0 | 5 | 10 | 15 | 20 | |
| e | 24 × 10 − ◊ = 228 | 12 | 14 | 16 | 18 | |
| f | ◊ ÷ 2 = $4^2$ + 3 | 35 | 36 | 37 | 38 | |
| g | (5 + ◊) × 10 = 25 × 3 | 1.5 | 2 | 2.5 | 3 | |

# Extended practice

A word puzzle can be made simpler by writing an equation. For example, guess my number: If you double it and add 3 the answer is 11. We can use ◊ for the number and write an equation: ◊ × 2 + 3 = 11. To solve the equation we can use "opposites": ◊ × 2 = 11 − 3. So, ◊ × 2 = 8. (Use "opposites" again.) ◊ = 8 ÷ 2 = 4

**1** Solve these puzzles by writing an equation.

a Guess my number. If you triple it and subtract 4, the answer is 11.

b Guess my number. If you multiply it by 10 and subtract 15, the answer is 19.

**2** Does your calculator use BODMAS? Use 1 + 2 × 4 to find out. Using BODMAS, the answer should be 1 + 8 = 9. Now try it on a calculator. If it gives the answer as 12, there is nothing wrong with the calculator, but think why it would do that.

a What is 10 + 2 × 4 − 2? _____

b What answer do you think a basic calculator will give?

c Try the problem on a calculator. What answer does it give? _____

d Investigate answers to the same number sentence by placing the brackets in different positions.

## Number challenge

**3** Using the digit "4" four times with any of the four operations, it is possible to come up with the answers 0, 1, 2, 3, 4, 5, 6, 7, 8 or 9. For example: (4 + 4 + 4) ÷ 4 = 3. Try to find the others. (There is more than one way of getting some of the answers.)

# UNIT 5: TOPIC 1
# Length

Our everyday units of length are kilometres (km), metres (m), centimetres (cm) and millimetres (mm). We can convert between them like this:

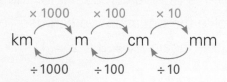

Remember! Only use a zero when it is needed.

## Guided practice

**1** Complete the table.

| | Kilometres | Metres |
|---|---|---|
| e.g. | 2 km | 2000 m |
| a | 4 km | |
| b | | 7000 m |
| c | | 19 000 m |
| d | 6 km | |
| e | 7.5 km | |
| f | | 3500 m |
| g | 4.25 km | |
| h | | 9750 m |

**2** Complete the table.

| | Metres | Centimetres |
|---|---|---|
| e.g. | 2 m | 200 cm |
| a | | 100 cm |
| b | 4 m | |
| c | 5.5 m | |
| d | | 250 cm |
| e | 7.1 m | |
| f | | 820 cm |
| g | 1.56 m | |
| h | | 75 cm |

**3** Complete the table.

| | Centimetres | Millimetres |
|---|---|---|
| e.g. | 2 cm | 20 mm |
| a | 5 cm | |
| b | 42 cm | |
| c | | 90 mm |
| d | 3.2 cm | |
| e | | 75 mm |
| f | | 125 mm |
| g | 12.4 cm | |
| h | | 99 mm |

**4** Which unit of length would you use for these?

a The length of this page _____

b The height of your table _____

c The length of an ant _____

d The length of a school hall _____

e The height of a door _____

f The length of a marathon race _____

## Independent practice

**1** Match **two** lengths to each object.

157 m    1.57 m    1570 m    157 mm    15.7 cm    0.157 km    1.57 km    157 cm    1.5 cm    15 mm

| | Object | 1st unit | 2nd unit |
|---|---|---|---|
| a | The length of a pencil | | |
| b | The height of a Year 6 student | | |
| c | The length of a finger nail | | |
| d | The distance around a school yard | | |
| e | The length of a bike ride | | |

**2** Measure each line. Write each length in three ways.

| | mm | cm and mm | cm |
|---|---|---|---|
| e.g. | 25 mm | 2 cm and 5 mm | 2.5 cm |
| a | | | |
| b | | | |
| c | | | |
| d | | | |

**3** Measure each object and its length in three ways.

| | Object | mm | cm and mm | cm (with decimal) |
|---|---|---|---|---|
| a | A pencil sharpener | | | |
| b | Your pencil | | | |
| c | An eraser | | | |
| d | A glue stick | | | |
| e | The width of this page | | | |

**4** Line B is 8 cm long.

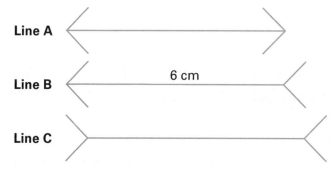

a  Estimate the lengths of the other two lines. (**Do not** measure the lengths.)

Line A estimate: _____     Line C estimate: _____

b  Measure Line A and Line C. Write the lengths.

Line A: _____     Line C: _____

**5** Line B here is 6 cm long.

a  Estimate (do not measure) the lengths of Lines A and C.

Line A estimate: _____     Line C estimate: _____

b  Now measure Lines A and C.

Line A: _____     Line C: _____

c  How did the arrows affect your estimates? _____

**6** Record the perimeter of each shape.

a  Perimeter = _____  b  Perimeter = _____  c  Perimeter = _____  d  Perimeter = _____

**7** Write about any shortcuts you used in question 6.

_____

_____

_____

# Extended practice

This is a list of estimated dinosaur lengths, measured from head to tail. Not all dinosaurs were gigantic. In fact, the shortest dinosaur has the longest name!

| Name | Length | Ranking (Longest to shortest) |
|---|---|---|
| Tyrannosaurus Rex | 12.8 m | |
| Iguanodon | 6800 mm | |
| Microraptor | 0.83 m | |
| Homalocephale | 290 cm | |
| Saltopus | 590 mm | |
| Puertasaurus | 3700 cm | |
| Dromiceiomimus | 3500 mm | |
| Micropachycephalosaurus | 50 cm | |

**1** Number the dinosaurs in order from **longest** to **shortest**.

**2** Name a modern animal that is about the same length as the smallest dinosaur.

_____

**3** Which dinosaur was about ten times longer than a dromiceiomimus?

_____

**4** If the longest dinosaur was lying on the ground, about how many Year 6 students could lay next to it, head to toe? _____

**5** What is the difference between your height and the length of a microraptor?

_____

**6** Draw a rectangle that has a perimeter of 68 mm.

# UNIT 5: TOPIC 2
## Area

Area is the surface of something. It is measured in squares. Depending on the size, we use square centimetres (cm²), square metres (m²), hectares (ha) or square kilometres (km²).

7 692 024 km²

## Guided practice

**1** This rectangle has one-centimetre squares drawn on it.

The area is _____ cm².

**2** This rectangle has one-centimetre squares drawn on part of it.

The area is _____ cm².

**3** This rectangle has marks every centimetre along two of the edges.

The area is _____ cm².

**4**

a How many centimetre squares would fit on the bottom row?

b How many rows would there be?

c What is the area?

**5** Write the area of each shape.

a

1 cm
2 cm

b
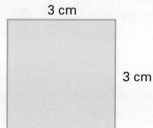
3 cm
3 cm

c

6 cm
3 cm

Area = _____      Area = _____      Area = _____

## Independent practice

**1** Find the length, width and area of each rectangle.

a  Length _____ cm

Width _____ cm

Area _____ cm²

b  Length _____ cm

Width _____ cm

Area _____ cm²

**2** These are the dimensions of three rooms. Calculate the area of each room.

*It is not always possible to draw shapes to their true size — because they wouldn't fit on the page!*

a  8 m, 5 m

b  9 m, 7 m

c  15 m, 10 m

Area = _____    Area = _____    Area = _____

**3** Scale is sometimes used in drawing plans. For these rooms, 1 cm on the plan represents 1 m in real life. Find the area of each room.

a 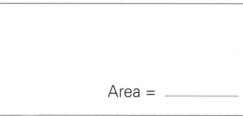 Area = _____

b Area = _____

c  Area = _____

4. To find the area of a rectangle, you can use this formula:
**A** (area) = **L** (length) × **W** (width).

Why **won't** the formula work for this shape? _____

5. Measure these shapes, then split them into rectangles to find the total area of each.

a

Area = _____

b

Area = _____

c

Area = _____

d
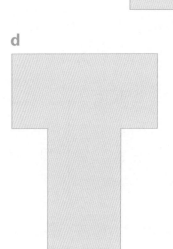
Area = _____

e

Area = _____

6. What is the area of a soccer field that is 100 m long by 50 m wide?

7. Two soccer fields side by side have an area of one hectare (ha). (1 ha = 10 000 m².) How many square metres are there in:

   a  2 ha? _____   b  4 ha? _____   c  5 ha? _____

8. An A4 page is 297 mm × 210 mm. Round to the nearest centimetre to find the approximate area of an A4 page.

# Extended practice

To find the area of a triangle, imagine it as half of a rectangle. The area of triangle ABC is half of rectangle ABCD. Half of 8 cm² = 4 cm².

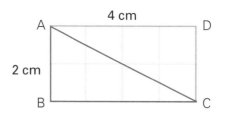

**1** Record the area of each shape.

**a** Area of rectangle ABCD = _____

Area of triangle ABC = _____

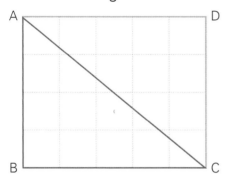

**b** Area of rectangle EFGH = _____

Area of triangle EFG = _____

**c** Area of rectangle IJKL = _____

Area of triangle JKL = _____

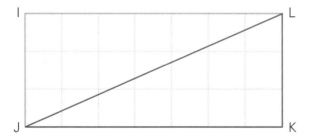

**d** Area of rectangle MNOP = _____

Area of triangle NOQ = _____

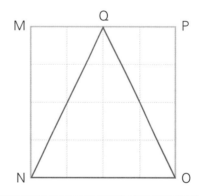

---

**2** Record the area of each triangle.

**a**      **b**      **c**

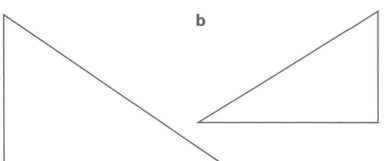 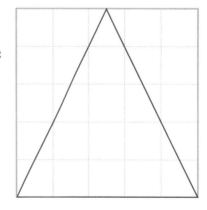

Area of triangle = _____    Area of triangle = _____    Area of triangle = _____

OXFORD UNIVERSITY PRESS

# UNIT 5: TOPIC 3
## Volume and capacity

**Volume** is the space something takes up. It is measured in cubes. This centimetre cube model has a volume of 4 cubic centimetres (4 cm³).

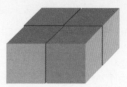

### Guided practice

**1** Write the volume of each centimetre cube model.

  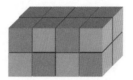    

a  Volume = _____ cm³
b  Volume = _____ cm³
c  Volume = _____ cm³
d  Volume = _____ cm³

**2** Fill in the gaps.

a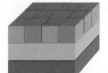
Top layer volume = _____ cm³
Number of layers _____
Volume of model = _____ cm³

b
Top layer volume = _____ cm³
Number of layers _____
Volume of model = _____ cm³

**Capacity** is the amount that can be poured into something. We use litres (L) and millilitres (mL) to show capacity. Large capacities (such as swimming pools) are measured in kilolitres (kL). We can convert between them like this:

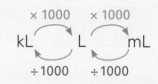

**3** Complete these tables.

a

| Kilolitres | Litres |
|---|---|
| e.g. 4 kL | 4000 L |
| 3 kL | |
| | 9000 L |
| | 3500 L |
| 6.25 kL | |

b

| Litres | Millilitres |
|---|---|
| e.g. 4 L | 4000 mL |
| | 2000 mL |
| 7 L | |
| 5.75 L | |
| | 4500 mL |

c

| Volume | Capacity |
|---|---|
| 1 cm³ = 1 mL | |
| e.g. 100 cm³ | 100 mL |
| | 500 mL |
| 225 cm³ | |
| | 1 L |
| 1750 cm³ | |

## Independent practice

**1**  
a How many centimetre cubes would you need to make this model? ☐

b What is its volume? ☐

**2** How do you know that the volume of this model is 12 cm³?

_____

_____

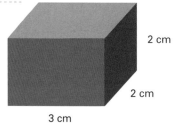

**3** Using **V** for volume, **L** for length, **W** for width and **H** for height, write a rule to tell someone how to find the volume of a rectangular prism.

_____

**4** Find the volume of each object.

a
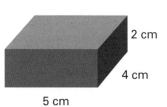

Volume: _____ cm³

b
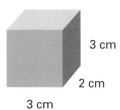

Volume: _____ cm³

c

6 cm

2 cm

4 cm

Volume: _____ cm³

d
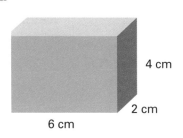

Volume: _____ cm³

e

Volume: _____ cm³

f
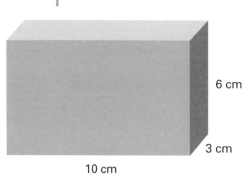

Volume: _____ cm³

**5** Order these containers by capacity from smallest to largest.

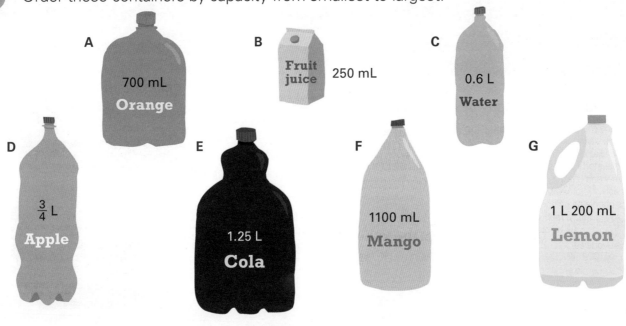

_____

_____

**6** Shade these jugs to show the level when the drinks have been poured in. Write the amount in millilitres (mL).

a  2 orange drinks

Amount: _____ mL

b  2 apple drinks

Amount: _____ mL

c  1 water and 1 orange drink

Amount: _____ mL

d  1 cola drink

Amount: _____ mL

e  3 fruit juice drinks

Amount: _____ mL

f  1 apple and 1 water drink

Amount: _____ mL

# Extended practice

**1** We would show the length and width of a driveway in metres, but how would we show its depth?

Would you use metres, centimetres or millimetres to show the depth of a driveway?

_____

Working-out space

**2** Ready-mixed concrete is sold by the cubic metre. How much concrete should be ordered for a path that is 30 m long, 3 m wide and 15 cm deep?

_____

**3** This activity is for finding the volume of a pebble.

You will need: a pebble (or similar), a small container of water, a bowl in which to place the small container, and a measuring jug.

*1 mL of water takes up the same space as 1 cm³.*

   **a** Place the smaller container into the bowl.

   **b** Carefully fill the smaller container with water up to the brim.

   **c** Gently place the pebble into the water.

   **d** Carefully remove the smaller container from the bowl, making sure that no more water spills from it.

   **e** Measure the amount of the water that spilled from the container when you placed the pebble into it.

   **f** Think about the connection between the amount of water that spilled over and the volume of the pebble.

_____

Write a few sentences about what you did. Include a sentence that states the volume of the pebble. Also say how you know what the volume is. (You may need to work this out on a separate piece of paper.)

_____

_____

_____

# UNIT 5: TOPIC 4
# Mass

The mass of something tells us how heavy it is.
Our everyday units of mass are tonnes (t), kilograms (kg) and grams (g).
For the mass of something very light, like a grain of salt, we use milligrams (mg).
Each unit of mass is 1000 times lighter than the next (heavier) unit.

## Guided practice

**1**

a

| Tonnes | Kilograms |
|---|---|
| e.g. 2 t | 2000 kg |
| 5 t | |
| | 7500 kg |
| | 1250 kg |
| 2.355 t | |
| | 995 kg |

b

| Kilograms | Grams |
|---|---|
| e.g. 2 kg | 2000 g |
| | 3500 g |
| 4.5 kg | |
| 0.85 kg | |
| | 250 g |
| 3.1 kg | |

c

| Grams | Milligrams |
|---|---|
| e.g. 4 g | 4000 mg |
| 5.5 g | |
| | 3750 mg |
| 1.1 g | |
| | 355 mg |
| 0.001 g | |

**2** What is something that would have its mass shown in:

a tonnes? _____   b kilograms? _____

c grams? _____   d milligrams? _____

**3** The mass of the box can be written as $1\frac{1}{2}$ kg, 1.5 kg or 1 kg 500g. Complete the table.

*The same mass can be shown in more than one way.*

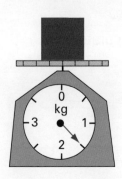

| | Kilograms and fractions | Kilograms and decimals | Kilograms and grams |
|---|---|---|---|
| a | $3\frac{1}{2}$ kg | | 3 kg 500 g |
| b | | 2.5 kg | |
| c | $3\frac{3}{4}$ kg | | |
| d | | 4.7 kg | |
| e | | | 1 kg 900 g |

## Independent practice

**1** We usually weigh heavier objects and lighter objects on different scales. Take note of the increments (markings) on each scale as you record the mass.

A 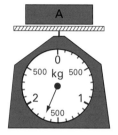   B 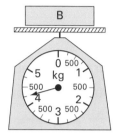   C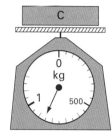

Mass: _____    Mass: _____    Mass: _____

**2** Look at the scales in question 1. Which would you use if you needed to have:

a  200 g of flour?  _____    b  $4\frac{1}{2}$ kg of potatoes?  _____

c  $2\frac{1}{4}$ kg of sand?  _____    d  750 g of apples?  _____

**3** 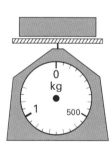  Draw a pointer on the scale to show a box with a mass of 925 g.

**4**

A    B    C    D

a  Order the trucks from lightest to heaviest load.  _____

b  Which two trucks are carrying a total mass closest to 5 t?  _____

c  Which two trucks are carrying a total mass closest to 6 t?  _____

OXFORD UNIVERSITY PRESS

**5** Finding the mass of a single sheet of paper would be difficult. Explain how you could use the information on this note pad to work out the mass of one sheet.

NOTE PAD
100 sheets

_____

_____

_____

**6** Sam receives a parcel from his grandmother. The total mass of the parcel is 1.85 kg. Write a possible mass for each item in the box. Make sure the total is 1.85 kg.

Working-out space

| Item | Mass |
| --- | --- |
| The packing box | |
| Pen set | |
| Shoes | |
| Set of postage stamps | |
| Pair of socks | |
| Packet of cookies | |

Which item do you think has the greatest mass?

**7** Each lift has a sign that shows the mass it can carry safely. Answer the following questions. Show your working out.

— LIFT —
Safe carrying capacity:
$\frac{1}{2}$ tonne (8 people)

a What does the lift company think is the average mass of a person?

b If the average mass of a Year 6 student is 40 kg, how many Year 6 students could the lift carry?

Working-out space

**8** In a 1-kg tray of four mangoes, none of them has the same mass. What might be the mass of each mango?

_____

_____

# Extended practice

**1** Scientists have shown that 1 mL of water has a mass of exactly 1 g. Could you prove that 1 mL of water has a mass of 1 g? In everyday life, it is very difficult to be accurate when weighing objects as light as one gram. Using a balance, try to prove that 50 mL of water has a mass of 50 g. Write a sentence or two about your findings.

_____

_____

**2** Sodium is part of salt, and we should not each too much of it.
This information shows the amount of sodium in some common foods.

| Type of food | Milligrams of sodium per 100 g | Normal serving size (g) | Milligrams of sodium per serve |
|---|---|---|---|
| Potato crisps | 1000 mg | 50 g | 500 mg |
| Hamburger | 440 mg | 200 g | 880 mg |
| Beef sausage | 790 mg | 70 g | 553 mg |
| Chicken breast | 43 mg | 160 g | 69 mg |
| Breakfast cereal | 480 mg | 30 g | 144 mg |
| Butter | 780 mg | 7 g | 55 mg |
| Yeast spread | 3000 mg | 6 g | 180 mg |
| White bread | 450 mg | 30 g (1 slice) | 135 mg |

We are not supposed to have more than about 2.3 g of sodium per day.

**a** Which type of food has 1 g sodium for every 100 g serving? _____

**b** What is the difference between the amount of sodium in 100 g of hamburger and 100 g of breakfast cereal?

_____

**c** Look at the normal serving sizes. How much sodium would Pete have if he ate a sandwich of two slices of white bread, yeast spread and butter?

_____

**d** If Helen were to eat one serving of each type of food in a day, by how much would she be over the recommended daily amount of sodium?

_____

# UNIT 5: TOPIC 5
## Timetables and timelines

A **timetable** is an easy-to-read list of what is going to happen. A **timeline** shows the order of things that have happened over a period of time.

Timetables use either 12-hour time or 24-hour time.

## Guided practice

**1** Fill in the missing times.

e.g.

a In the afternoon

b In the evening

am/pm time: 5:16 am     _____     _____

24-hour time: 0516     _____     _____

c

d Late at night

e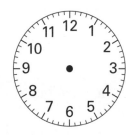

am/pm time: 2:42 am     _____     _____

24-hour time: _____     _____     2222

f In the morning

g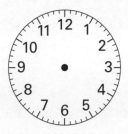

h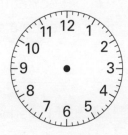

am/pm time: _____     10:35 am     _____

24-hour time: _____     _____     2359

## Independent practice

**1** How long does it take for Train 8219 to get from Melbourne to Geelong?

_____

**2** Which train takes the shortest time to get from Melbourne to Geelong?

_____

**3** What is the difference between the times of the shortest and longest journeys from Melbourne to Geelong?

_____

**4** On which train is it possible to buy a drink?

_____

| | SATURDAY | | | | |
|---|---|---|---|---|---|
| SERVICE NO. | 8215 | 8219 | 8221 | 8225 | 8227 |
| Train/Coach | TRAIN | TRAIN | TRAIN | TRAIN | TRAIN |
| Seating/Catering | ★ ☕ | | | | |
| MELBOURNE | | IC | | | |
| (Southern Cross) dep. | 13:00 | 14:00 | 15:00 | 16:00 | |
| North Melbourne | | | | | |
| Footscray | 13:08u | 14:08u | 15:08u | 16:08u | |
| Newport | | | | | |
| WERRIBEE | | | 14:26u | | 16:26u |
| Little River | | | 14:34 | | 16:34 |
| Lara | 13:42 | | 14:40 | 15:35 | 16:40 |
| Corio | | | 14:44 | | 16:44 |
| North Shore | | | 14:46 | | 16:46 |
| North Geelong | 13:50 | | 14:50 | 15:43 | 16:50 |
| GEELONG arr. | 13:56 | | 14:54 | 15:47 | 16:54 |

**Legend**
★ First Class available.   ☕ Catering available.   arr. – Arrive.   dep. – Depart.
u – Stops to pick up passengers only.   IC – Inter-City.   W – To Warrnambool.
▓ Peak service.   ░ Reservation required on these services.

**5** Why could you **not** travel from Southern Cross Station to Footscray on any of the trains?

_____

**6** If you were going to Geelong on Train 8219 and wanted get off in Lara to meet a friend, how long would you have to wait for the next train? _____

**7** Train 8227 leaves Southern Cross at 4:30 pm. It has the same travelling time and stops as Train 8225. Fill in the blanks on the timetable. Use 24-hour time.

**8** Use the information about space travel to complete the timeline. Make sure to take note of the scale, write the year for each mark on the scale, and draw the arrows to the appropriate places on the timeline.

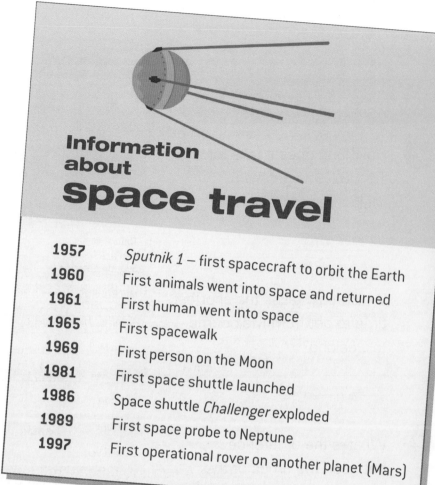

## Information about space travel

| 1957 | Sputnik 1 – first spacecraft to orbit the Earth |
| 1960 | First animals went into space and returned |
| 1961 | First human went into space |
| 1965 | First spacewalk |
| 1969 | First person on the Moon |
| 1981 | First space shuttle launched |
| 1986 | Space shuttle *Challenger* exploded |
| 1989 | First space probe to Neptune |
| 1997 | First operational rover on another planet (Mars) |

## A timeline of space exploration

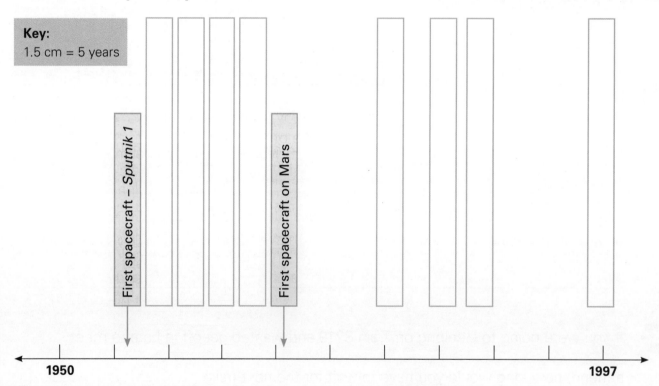

**Key:** 1.5 cm = 5 years

**9** According to the timeline, in which year did the first spacecraft land on Mars?

Which do you find easier to understand: the list of events or the timeline? Why?

# Extended practice

**1** There are three buses a day from Small Town to Big Town.

|       | Departs Small Town | Arrives Big Town |
|-------|--------------------|------------------|
| Bus A | 0752               | 1043             |
| Bus B | 1114               | 1408             |
| Bus C | 1526               | 1829             |

a About how many hours does the bus take to get from Small Town to Big Town?

b How long is the journey on Bus A?

c How much longer is the journey on Bus C than Bus B?

**2** Look at the timetable in question 1.

a Write the departure time for Bus C in am/pm time.

b Draw the time on the analogue clock.

c If you took Bus B to Big Town and the person meeting you did not arrive until 2:30 pm, how long would you have to wait?

**3** Each bus waits at Big Town for 85 minutes before starting the return journey. Each return journey takes 2 hours and 59 minutes. Complete the timetable for the journeys from Big Town to Small Town using 24-hour time.

|       | Departs Big Town | Arrives Small Town |
|-------|------------------|--------------------|
| Bus A |                  |                    |
| Bus B |                  |                    |
| Bus C |                  |                    |

# UNIT 6: TOPIC 1
## 2D shapes

A polygon is a 2D shape with straight sides. *Regular* polygons have sides and angles that are the same size. *Irregular* polygons do not.

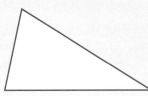

Is a circle a polygon?

## Guided practice

**1** Name these shapes and label them as either *regular* or *irregular*.

| | | Name of shape | Regular or irregular? |
|---|---|---|---|
| e.g. | △ | triangle | regular |
| a | ⬡ | | |
| b | ⬠ | | |
| c | ▢ | | |
| d | ⬠ | | |
| e | ⬢ | | |
| f | ⬠ | | |
| g | ◁ | | |
| h | ⬠ | | |

**2** Circle the sentence that best describes this polygon.

- It has five sides.
- It has five angles that are the same size.
- It has five equal sides and some of the angles are the same size.
- It has five equal sides and five angles that are the same size.
- It has five equal sides but no angles that are the same size.

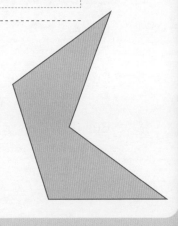

## Independent practice

**1** Write the type of triangle and two of its properties.

| | Triangle | Type and properties |
|---|---|---|
| | | Scalene. The sides are different lengths. It has an obtuse angle. |
| a | | |
| b | | |
| c | | |
| d | | |
| e | | |

**2** Identify and describe each shape.

| | Quadrilateral | Special name | Description |
|---|---|---|---|
| | | Irregular quadrilateral | The four sides are all different lengths. The angles are different sizes. It has a reflex angle. |
| a | | | |
| b | | | |
| c | | | |
| d | | | |
| e | | | |

> Every square is a rhombus, but not every rhombus is a square.

**3** Write about the similarities and differences between each pair of shapes.

| Shapes | Similarities | Differences |
|---|---|---|
| △ △ | The angles in both triangles are acute. Each triangle has at least two angles that are the same size. | One triangle has two sides that are the same length. The other has all three sides the same length. |
| a | | |
| b | | |
| c | | |
| d | | |
| e | | |
| f | | |
| g | | |
| h | | |
| i | | |
| j | | |
| k | | |

# Extended practice

**Word bank**

semi-circle   circumference
quadrant   radius
diameter   sector

**1** Fill the gaps. The arrow is pointing to a:

a _____   b _____   c _____

**2** Fill the gaps. The shaded part is a:

a _____   b _____   c _____

  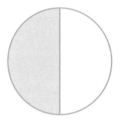

**3** Draw a circle with a diameter of 12 cm inside the square. Use the dot (A) as the centre of the circle.

**4** Draw the diagonals on the square. How many triangles are there?

# UNIT 6: TOPIC 2
## 3D shapes

A pyramid has one base. The shape of the base gives the pyramid its name.

A prism has two bases (ends). The shape of the bases gives the prism its name.

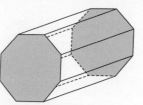
Octagonal prism

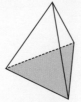
Triangular pyramid

*A prism doesn't always sit on its base.*

## Guided practice

**1** Use the shapes of the bases to identify these 3D shapes.

a

b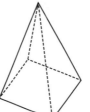

c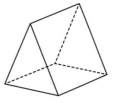

_____   _____   _____

d

e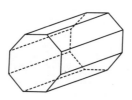

f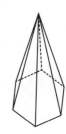

_____   _____   _____

g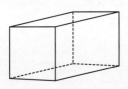

h

i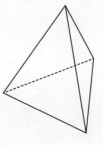

_____   _____   _____

## Independent practice

1. These are the nets for which 3D shapes?

    A _____

    B _____

2. Trace or copy the nets and make the 3D shapes. Think about adding tabs to help you glue the faces together.

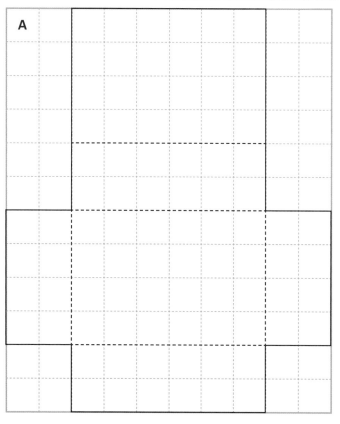

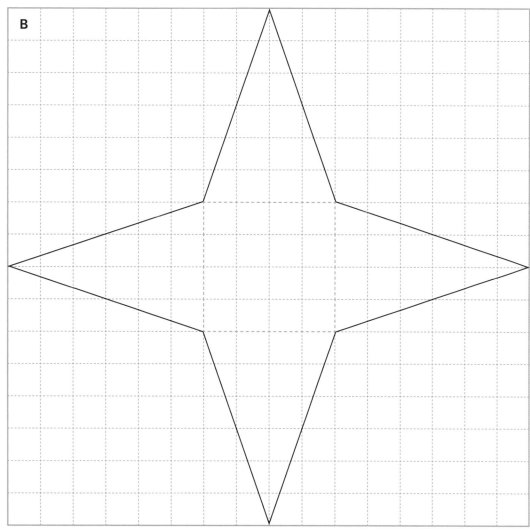

3. Practise drawing these 3D shapes.

It's OK to make mistakes. It means you're learning!

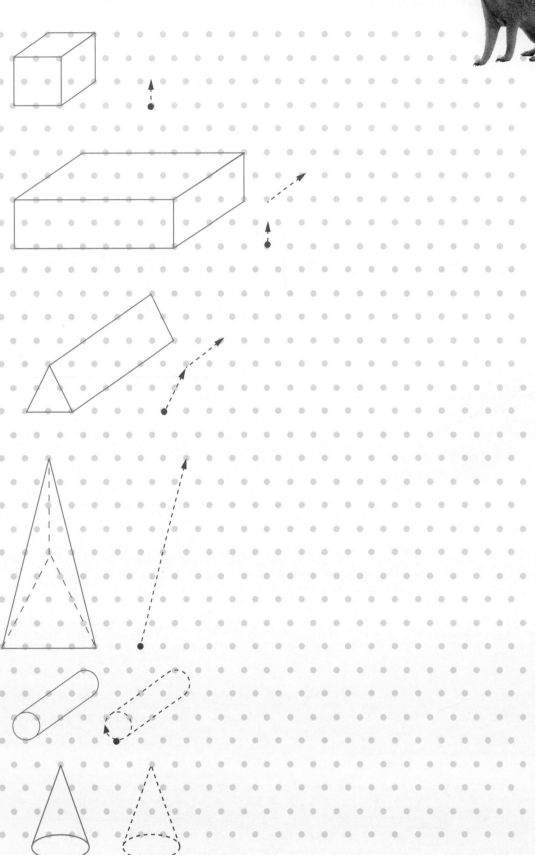

# Extended practice

**1** Euler's law says that if you add the number of faces and vertices on a 3D shape, then take away the number of edges, the answer is always 2. Test the law on these objects.

| | Object | Name | Number of faces | Number of vertices | Number of edges | Does Euler's Law work? |
|---|---|---|---|---|---|---|
| a | | Rectangular prism | 6 | 8 | 12 | Yes |
| b | | | | | | |
| c | | | | | | |
| d | | | | | | |
| e | | | | | | |
| f | | | | | | |
| g | | | | | | |
| h | | | | | | |

# UNIT 7: TOPIC 1
# Angles

An angle has two arms and a vertex. A protractor is used to measure angles. The unit of measurement is called a degree (°).

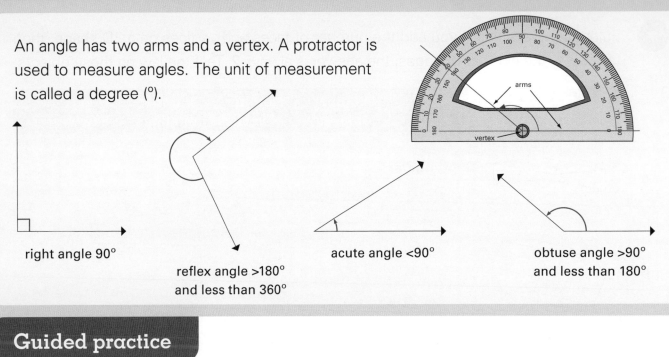

right angle 90°

reflex angle >180° and less than 360°

acute angle <90°

obtuse angle >90° and less than 180°

## Guided practice

**1** Write the size and type of each angle.

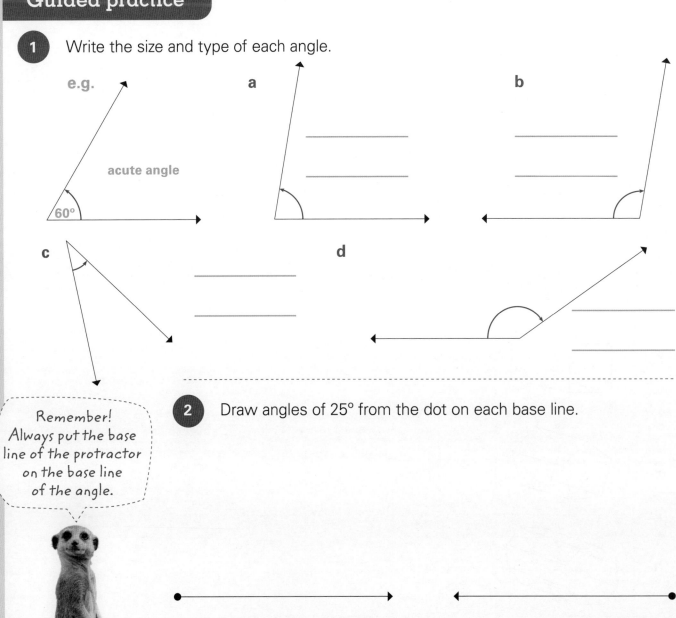

e.g. acute angle 60°

**2** Draw angles of 25° from the dot on each base line.

Remember! Always put the base line of the protractor on the base line of the angle.

# Independent practice

**1** Write the size of each angle inside the arc.

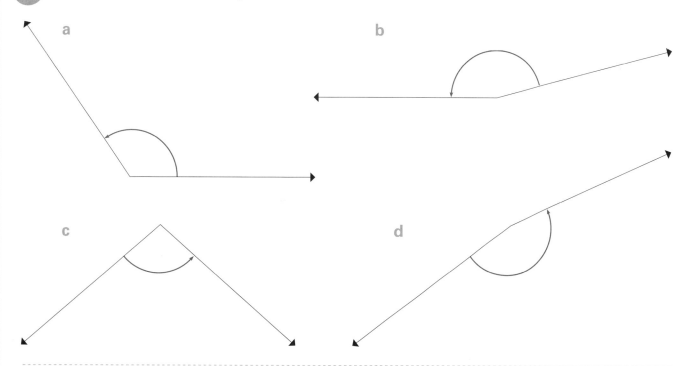

**2** How do you know the size of the reflex angle is 320°?

_____

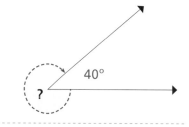

**3** Write the size of each reflex angle.

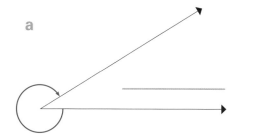

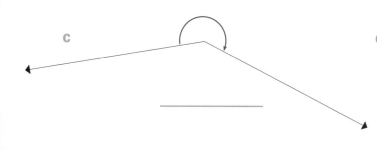

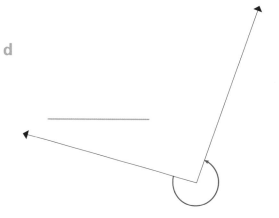

**4** Calculate the size of each unknown angle without using a protractor.

Remember! A straight angle is 180° and a revolution is 360°.

e.g. ? = 100°
(180 − 80)

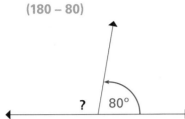

a ? = _____

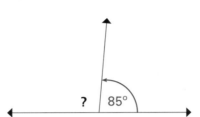

b ? = _____

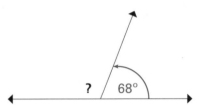

c ? = _____

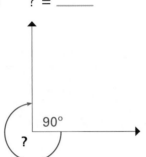

d ? = _____

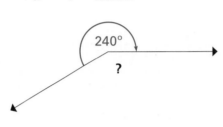

e ? = _____

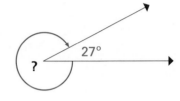

f ? = _____

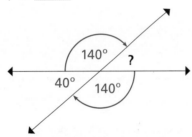

g ? = _____

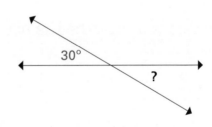

h ? = _____

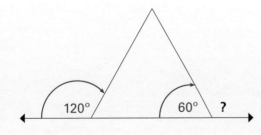

i ? = _____

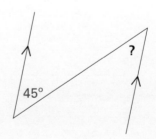

j ? = _____

# Extended practice

**1** Calculate the sizes of the angles.

**a**  a = \_\_\_\_\_
     b = \_\_\_\_\_

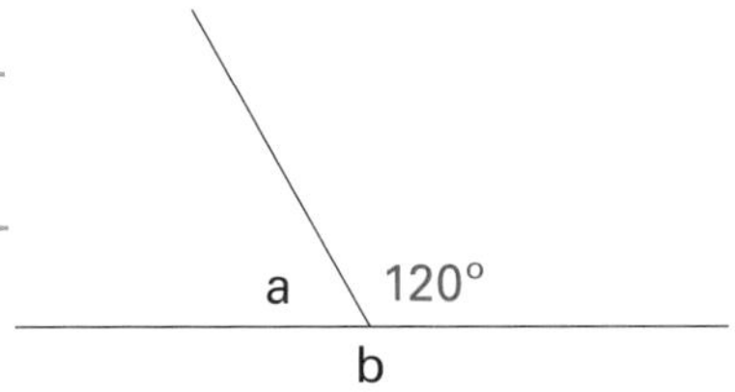

**b**  a = \_\_\_\_\_
     b = \_\_\_\_\_
     c = \_\_\_\_\_

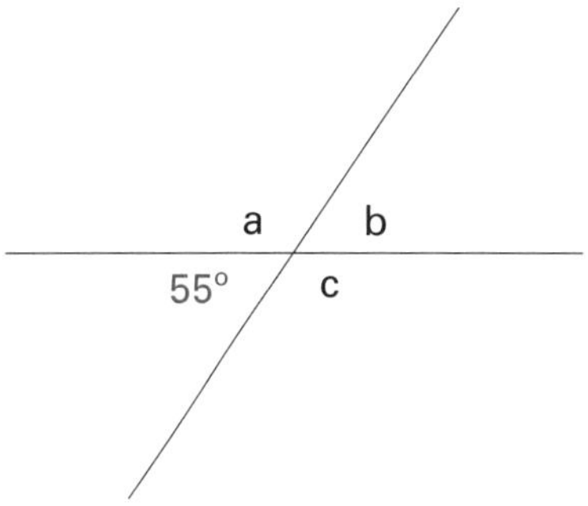

**c**  a = \_\_\_\_\_
     b = \_\_\_\_\_
     c = \_\_\_\_\_

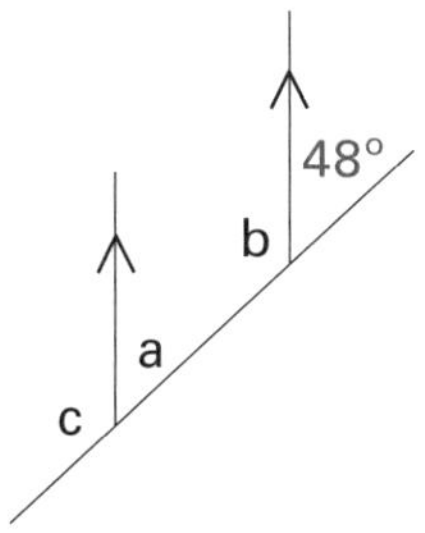

**d**  a = \_\_\_\_\_

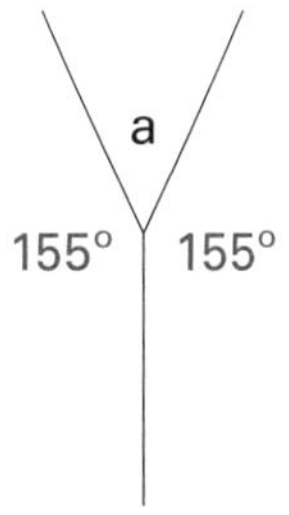

**2** From the information shown, calculate the sizes of the marked angles.

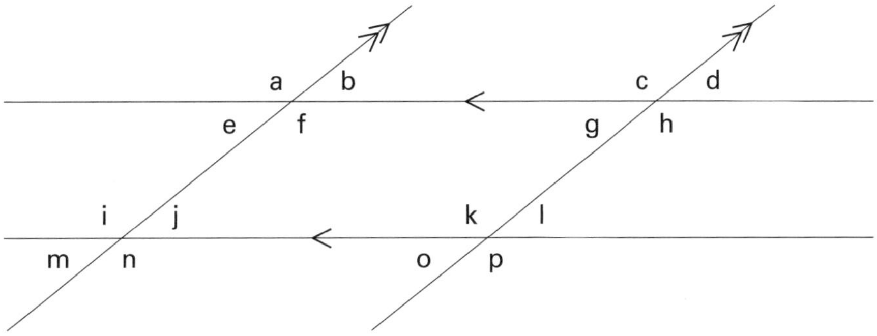

| a = 142° | b = \_\_\_\_\_ | c = \_\_\_\_\_ | d = \_\_\_\_\_ |
| e = \_\_\_\_\_ | f = \_\_\_\_\_ | g = \_\_\_\_\_ | h = \_\_\_\_\_ |
| i = \_\_\_\_\_ | j = \_\_\_\_\_ | k = \_\_\_\_\_ | l = \_\_\_\_\_ |
| m = \_\_\_\_\_ | n = \_\_\_\_\_ | o = \_\_\_\_\_ | p = \_\_\_\_\_ |

**3** Use the base line to draw a parallelogram. The second arm of the angle at Point A is 75° up from line AB. It is 5 cm long.

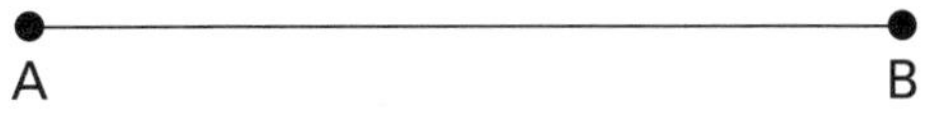

# UNIT 8: TOPIC 1
# Transformations

Patterns can be made by transforming a shape. This could be by:

Translation (sliding it)

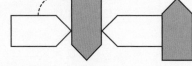
Rotation (turning it)

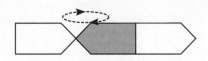
Reflection (flipping it over)

## Guided practice

**1** What method of transformation has been used?

*Transforming is another word for changing.*

a  _____

b  _____

c  _____

**2** a Reflect the triangle.

b Rotate the pentagon.

c Translate the parallelogram.

# Independent practice

Patterns can be made by transforming shapes *horizontally* or *vertically*.

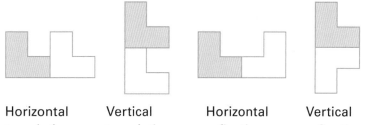

Horizontal translation   Vertical translation   Horizontal reflection   Vertical reflection

**1** Describe the patterns.

| Pattern | Description |
|---|---|
| (trapezium) | The trapezium has been reflected vertically. |
| a (hexagons) | |
| b (triangles) | |
| c (hexagons) | |
| d (pentagons) | |
| e (triangles) | |
| f (arrows) | |

**2** Continue the pattern and describe the way it grows.

3) Complete and describe these patterns. (There is no need to include the colours in your description.)

a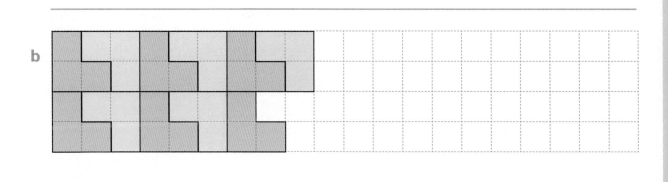

___

b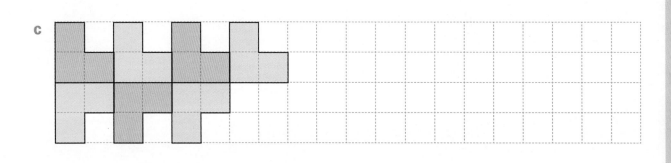

___

c

___

4) Design a transformation pattern. Use this shape or make up one of your own.

## Extended practice

**1**

For these activities you will need a computer with Microsoft Word, or similar.

- **a** Choose an object that you find interesting. Place it on the page and fill it with the colour of your choice.
- **b** Copy and paste the object on the same point and rotate it through an angle (say, 30°).
- **c** Repeat this process until you have made a design you are happy with.
- **d** Save your design and, if you have permission, print it.

---

**2**

- **a** Open a new Word document.
- **b** Choose a basic shape (such as the trapezium in this pattern).
- **c** Draw a shape at the top of the page by clicking and dragging.
- **d** Copy the shape and paste it next to the original.
- **e** Reflect the second shape vertically and position it next to the first.
- **f** Select **both** objects. Copy and paste them so that they join the original pair.
- **g** Look for shortcuts to continue your pattern.
- **h** After placing 8 shapes, group them, copy them and paste them below the first row.
- **i** Reflect the second row vertically and horizontally.
- **j** Continue until you have 8 or more rows.

# UNIT 8: TOPIC 2
## The Cartesian coordinate system

The Cartesian plane was named after René Descartes. It is split into four quadrants (or quarters). The *x*-axis and the *y*-axis meet in the middle at the **origin point**.

Numbers to the left of the origin point are negative. Numbers below the origin point are negative. Points are named by pairs of numbers called *ordered pairs*. Always read the number on the *x*-axis first.

*Who was René Descartes? Why are Cartesian coordinates named after him? Try to find out!*

## Guided practice

**1** The blue triangle is at point (−4,5). What is at (4,−5)?

_____

**2** Write the coordinate points for:

  **a** the green circle

  **b** the pink circle

**3** If you drew a straight line from (−8,−6) to (4,−5), which two shapes would it join? _____

**4** True or false? If you drew a line from the centres of the two triangles, it would pass through the origin point. _____

**5** **a** Draw a line from (0,0) to (4,−4).

   **b** Through which other points does it pass? _____

   **c** Draw a small square at (−7,2).

   **d** Draw a smiley face at (9,−3).

## Independent practice

**1** Give the coordinate points for the:

a  yellow dot:

b  green dot:

c  red dot:

d  blue dot:

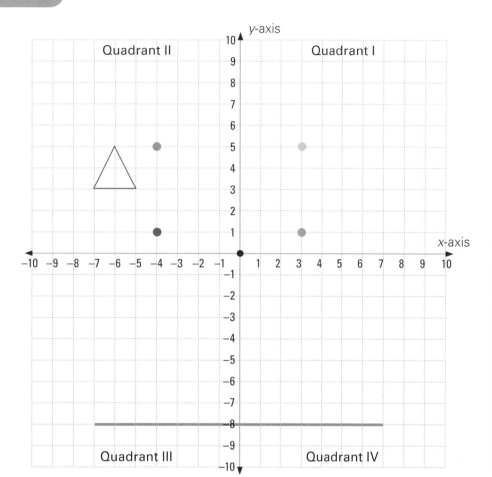

If you draw a line on the Cartesian plane, you can use an arrow to show the points that are joined. You would plot the blue line by writing (–7,–8) → (7,–8).

**2** Finish the ordered pairs for the red triangle in question 1:

(–7, 3) → (–5, 3) → ☐ → ☐

**3** Show how you could join points that would draw a rectangle between the four dots in question 1. Remember to "close" the rectangle.

**4** a  Draw a simple 2D shape in Quadrant IV.

b  Write the ordered pairs that would draw your shape.

**5** Draw a face by following the ordered pairs below.

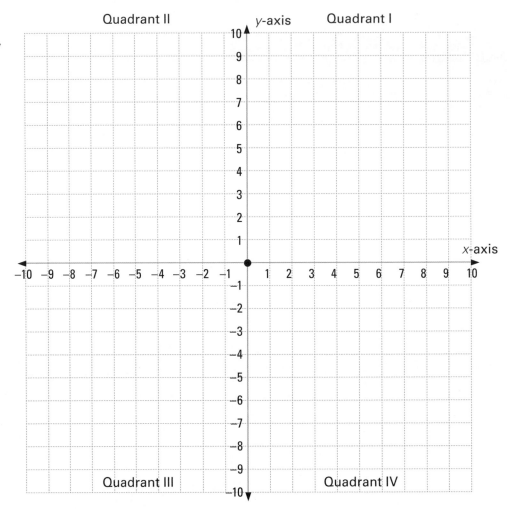

a  (4,6) → (8,6) → (8,10) → (4,10) → (4,6)

b  (5,8) → (5,7) → (7,7) → (7,8)

c  Draw dots at (5,9), at (6,8) and at (7,9).

---

**6**  a  Plot and record the points that would draw a large hexagon in Quadrant II.

_____

b  Plot and record the points that would draw a large pentagon in Quadrant III.

_____

c  Plot and record the points that would draw a large octagon in Quadrant IV.

_____

d  In Quadrant I, create a simple picture using straight lines. Write the coordinates that someone could follow to draw the same picture.

_____

# Extended practice

**1** Will's bedroom always seemed to be untidy. One day his mother was so tired of it that she hid everything that Will had left on the floor. When he got home all he found was a chair in the centre of the room.

Will's mother told him to sit on the chair, and she gave him a copy of the number plane. She also gave him a list of the coordinate points for everything he had left on the floor. She said he could have them back if he correctly plotted them on the number plane.

Plot all the items Will had left around the room. Use a letter for each item (or draw small illustrations at each coordinate point).

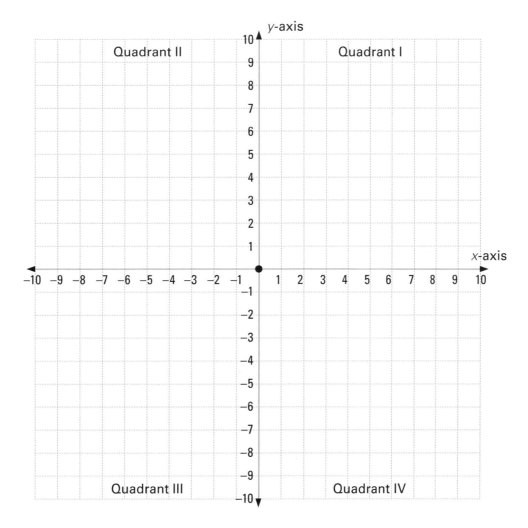

| A | music player: (2,5) | B | smart phone: (−3,3) | C | watch: (1,−2) |
| D | right shoe: (−3,6) | E | left shoe: (3,−6) | F | pencil case: (−4,4) |
| G | calculator: (−6,4) | H | comb: (4,2) | I | deodorant: (−3,−2) |
| J | video game: (5,7) | K | wallet: (5,−3) | L | bag: (−5,1) |
| M | shirt: (−2,−7) | N | jeans: (3,−2) | O | sock: (−2,1) |

# UNIT 9: TOPIC 1
## Collecting, representing and interpreting data

A common way to represent data is on a graph. There are several types of graphs. The type of graph used depends on what is being represented.

### Guided practice

**1** These are horizontal and a vertical bar graphs.

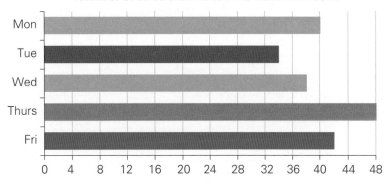

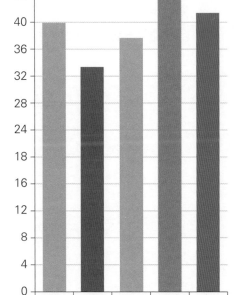

a  How many birds came on Friday? ☐

b  By how many was Tuesday's total less than Monday's total? ☐

**2** This is a dot plot.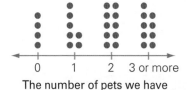

What was the most common number of pets that people in the class have? ☐

**3** This is a line graph.

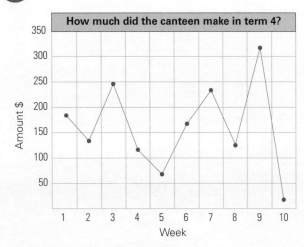

Estimate the amount the canteen made in Week 9. _____

**4** This is a pictograph.

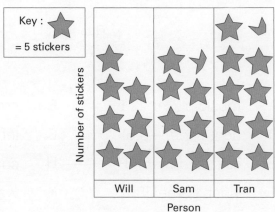

How many more stickers does Tran have than Sam? _____

## Independent practice

**1 a** Use the information in the frequency table to make a pictograph. Take note of the key.

**b** What is the difference between the total of the two favourite colours and the total of the two least favourite colours?

**Frequency table: Favourite colours for Year 6**

| Colour | Red | Yellow | Blue | Green | Purple |
|---|---|---|---|---|---|
| Frequency | 24 | 10 | 26 | 19 | 13 |

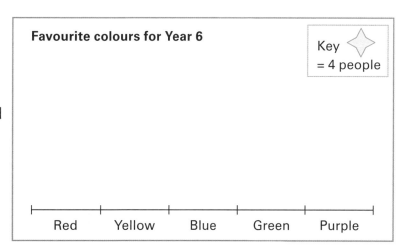

**2** Add your favourite colour and the favourite colour of five other people to the information in question 1, then rewrite the frequency table.

**Frequency table: Favourite colours for Year 6**

| Colour | Red | Yellow | Blue | Green | Purple |
|---|---|---|---|---|---|
| Frequency | | | | | |

**3 a** Transfer the information from question 2 onto a bar graph. Decide on a suitable scale.

**b** Is it better to use a pictograph or a bar graph to present this type of information? Give a reason for your answer.

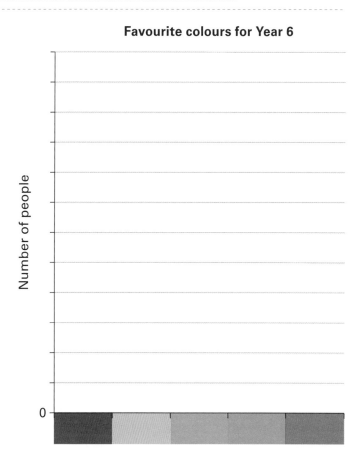

**4** This data shows hourly temperatures at a ski resort.

| Time | 0700 | 0800 | 0900 | 1000 | 1100 | 1200 | 1300 | 1400 | 1500 | 1600 | 1700 |
|---|---|---|---|---|---|---|---|---|---|---|---|
| Temperature | 0°C | 1°C | 2°C | 3°C | 7°C | 8°C | 8°C | 6°C | 5°C | 2°C | 1°C |

a  Show the information on a line graph. Remember to label the graph.

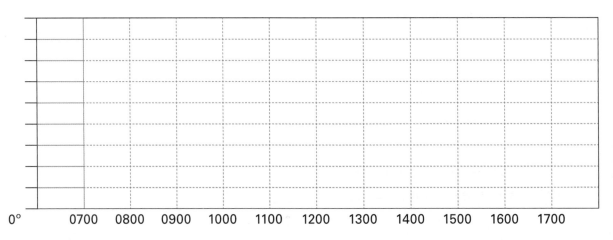

b  Write two statements about the information shown on the graph.

_____

_____

**5** Collect data about the hair colour of students in your class. Organise it on this frequency table.

| Hair type | Dark | | Fair | | Medium | | Other | |
|---|---|---|---|---|---|---|---|---|
| | Long | Short | Long | Short | Long | Short | Long | Short |
| **Number** | | | | | | | | |

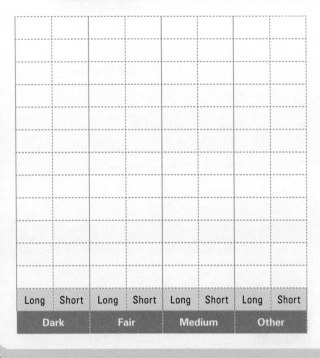

a  Choose an appropriate graph type to represent your data. The grid may help you.

*Make sure your graph is easy to interpret.*

b  Write two statements about the information shown on the graph.

_____

_____

_____

# Extended practice

**1** The circle graphs on this page show five of the Top 100 names for baby girls and boys in the first decade of this century.

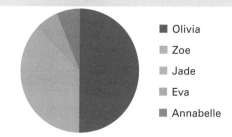

    **a** The combined total of which three names was about the same as the number of babies called Zoe?

    _____

    **b** Approximately half of the circle is used for the name Olivia. About what fraction is used for Eva?

**2** The same data about baby names is shown on the bar graph.

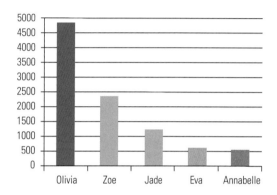

    **a** What information is shown on this graph that is not shown on the sector (circle) graph?

    _____

  **b** The number of babies named Jade was 1234. There were 14 more babies called Eva than Annabelle. Estimate the numbers of babies named Eva and Annabelle.

**3** Apart from the way that the graphs are shaded, what are the similarities or differences between the circle graphs for the popular names for girls and boys? Write some statements of finding.

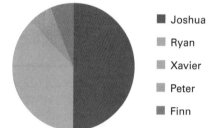

_____

_____

**4** This bar graph represents the same data as the circle graph in question 3.

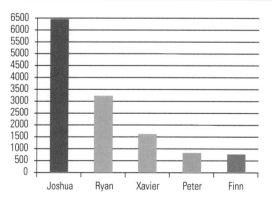

    **a** The number of babies named Xavier can be rounded to 1600. What is your estimate of the exact number?

    _____

  **b** Comment on the differences between the most popular names for girls and boys from the data shown on the two bar graphs.

_____

# UNIT 9: TOPIC 2
## Data in the media

Data that you collect yourself is called *primary data*. Some graphs are based on *secondary data*. This is when people use someone else's data.

## Guided practice

**1** Are the following likely to be based on *primary* or *secondary* data?

   **a** You make a graph about the Top 10 holiday places after reading a magazine article.

   **b** You make a graph about the favourite food of the people in your group.

When TV shows or newspapers collect data about what people think, they cannot ask everybody. They do a *sample* survey. If you get the views of everyone, it is called a *census* survey.

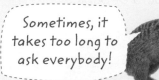

**2** Read the following scenarios. Were the surveys likely to be *sample* or *census*?

   **a** A phone company wanted to know what size of phone people prefer.

   **b** A Year 2 class did a graph about their favourite colours.

   **c** A principal wanted to know what parents thought about a new school uniform.

   **d** A newspaper boss wanted to know what local people thought about having a new skate park.

**3** A Year 6 teacher wants to know if the class would like to work on their science projects at lunchtime or at home. She surveys a group of eight of the 24 students.

   **a** Would the collected data be *primary* or *secondary*?

   **b** Was this a *sample* or a *census* survey?

## Independent practice

**1** The principal of the school in which the Year 6 teacher did a survey about the science project (see page 130) made a graph and published it in the school newsletter. Was the principal's graph based on *primary* or *secondary* data?

_____

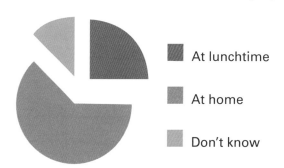

**When should Year 6 work on the science project?**

- At lunchtime
- At home
- Don't know

**2** Look at the information in the graph above. The teacher had surveyed eight of the 24 students in the class.

a  How many of the eight students wanted to do the project at home? ☐

b  How many answered *I don't know*? ☐

**3** The principal wrote in the newsletter: "The majority of the students surveyed prefer to work on their project at home." Is this a true statement of finding? Give a reason for your answer.

_____

**4** A parent contacted the principal and said, "If all 24 of the students had been surveyed, the result would have been different." Is this statement true? Underline one response:
- It is true
- It is not true
- It could be true.

**5** The boss of the town's newspaper saw the school newsletter and wrote a newspaper headline based on the story:

> Recent survey shows that most students in our town want more homework.

Underline one response. The headline is based on fact …

- and is partly true
- and is definitely true
- but has no truth to it
- and could be true.

Did you know that Australians spend more money on video games per person per year than Americans do? Kids from the UK spend the most! The following data is true, but the newspaper headline is fake.

**6**

a According to the graph, approximately how much per year does each Australian spend on video games? ☐

b In how many of the Top 10 countries does each person spend over $100 a year on video games? ☐

c According to the graph, approximately how much per year does each Italian spend on video games? ☐

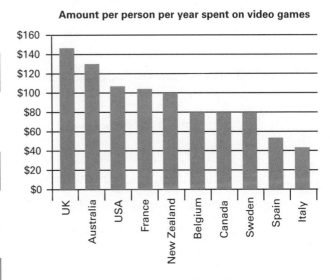

**7** A TV presenter in the UK was shown the newspaper article about video games. He told viewers, "Video games are making our children lazy. We should stop this NOW. Call us and give your opinion!" Later, the presenter told viewers that his research showed that 90 per cent of parents want children to be banned from playing video games.

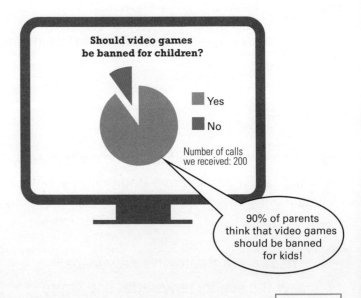

a Did the TV presenter conduct a census survey or a sample survey? ☐

b Do you think the presenter's statement was a fair one? Give a reason for your answer. _____

c How many parents actually answered "Yes" to the presenter's question? ☐

## Extended practice

> New survey shows that most people want a new fast-food restaurant.

**1** A newspaper in a town with a population of about 8000 printed an article about a new fast-food restaurant being built next to the local high school. The article was based on a survey carried out by a group of students. 100 people had been asked about the new restaurant.

    **a** Did the students carry out a sample or a census survey?

    **b** Did the newspaper use primary or secondary data?

**2**   **a** The majority of people said that the fast-food restaurant was a good idea. How many people might this have been?

    **b** What is the largest percentage of those surveyed who might have objected?

    **c** The newspaper did not mention anything about who had been surveyed. Some people complained to the newspaper editor about this. What do you think the main complaint was?

_____

**3** The following week the newspaper published an apology. It wrote that the 100 people who had been asked were students from the high school and that 97 of them thought the fast-food restaurant was a good idea.

    **a** In what way were the survey results not a fair reflection of public opinion?

_____

    **b** What would have been a fairer way to carry out the survey?

_____

    **c** Look again at the newspaper headline. Comment on the level of truth in the statement. _____

**4** The survey question was, "The restaurant has promised to give away 50 free burgers every week. Do you want a fast-food restaurant next to the school?"

    **a** Why was including the part about the free burgers not appropriate?

_____

    **b** Write a survey question that would be appropriate to find out people's opinions about the new restaurant.

_____

# UNIT 9: TOPIC 3
## Range, mode, median and mean

Range, mode, median and mean are all part of working out averages. We can use these weekly test scores to show range, mode, median and mean.

| Week | 1 | 2 | 3 | 4 | 5 |
|---|---|---|---|---|---|
| Score | 8 | 4 | 3 | 2 | 8 |

tall    average height    short

## Guided practice

### Range

The range is the difference between the highest and lowest in a set of numbers. In the test scores above, the range is 8 – 2 = 6. So, the range is 6.

**1** Find the range in these sets of numbers.

    **a** 22%, 16%, 64%, 80%, 31% _____

    **b** 75, 81, 150, 110, 95 _____

### Mode

The mode is the number that occurs most often. In the test scores above, 8 occurs more often than the others, so 8 is the mode. The mode is sometimes used to describe the average.

**2** Find the mode in these sets of numbers.

    **a** 35%, 34%, 44%, 35%, 31% _____

    **b** 75, 76, 75, 76, 76 _____

### Median

The median is the number in the middle of an ordered set of numbers. In order, the scores above are 2, 3, 4, 8, 8. The middle number is 4, so 4 is the median score. The median can also be used to describe the average.

**3** Find the median in these sets of numbers.

    **a** 76%, 44%, 24%, 15%, 71% _____

    **b** 15, 16, 25, 26, 15 _____

### Mean

To find the mean, add up all the numbers in the set and divide by however many numbers there are. In the scores above, the total of the five numbers is 25. Then divide 25 by 5, which equals 6. So the mean score is 6. The mean is most often used to describe average.

**4** Find the mean in these sets of numbers.

    **a** 36%, 20%, 36%, 24%, 34% _____

    **b** 15, 16, 14, 13, 17 _____

## Independent practice

**1** This table shows the minimum temperatures over a four-week period. Order the numbers for each week from lowest to highest. Then find the range, mode, median and mean temperatures.

| Week | Seven-day minimum temperatures | Order | Range | Mode | Median | Mean |
|---|---|---|---|---|---|---|
| 1 | 3°, 6°, 7°, 9°, 7°, 8°, 2° | | | | | |
| 2 | 1°, 3°, 2°, 9°, 7°, 7°, 6° | | | | | |
| 3 | 9°, 6°, 8°, 8°, 10°, 7°, 8° | | | | | |
| 4 | 10°, 9°, 10°, 8°, 7°, 3°, 2° | | | | | |

**2** This set of six numbers has no middle number.

**5, 6, 7, 8, 9, 10**

To find the median if there is an even-numbered set of numbers, add the *two* middle numbers and then divide the total by 2.

The two middle numbers in the set above are 7 and 8, which total 15. So, $15 \div 2 = 7\frac{1}{2}$ or 7.5. This is the median number.

Find the range and median in these sets of numbers.

*The median number does not always appear in the set of numbers.*

| Number set | Order | Range | Median |
|---|---|---|---|
| 8, 2, 6, 4, 10 | | | |
| 25, 14, 17, 12, 6, 4 | | | |
| 12, 8, 2, 6, 2, 5, 21 | | | |
| 82, 23, 3, 8, 15, 3, 16, 2 | | | |

**3** This set of numbers does not have a mode because no number occurs more often than any other. Sometimes it is not possible to show the mode.

**25, 16, 11, 17, 19, 16**

Find the mode and mean. If there is no mode write "none".

| Number set | Mode | Mean |
|---|---|---|
| 8, 2, 6, 4, 10 | | |
| 25, 14, 17, 12, 6, 4 | | |
| 12, 8, 2, 6, 2, 5, 21 | | |
| 82, 23, 3, 8, 15, 3, 16, 2 | | |

**4** This graph compares the average hours of sunshine per day in London and Sydney.

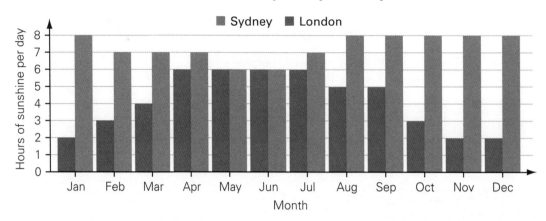

a Look at the graph. Without doing any calculations, estimate the daily average hours of sunshine for the whole year.

|  | Sydney | London |
|---|---|---|
| Mode |  |  |
| Mean |  |  |

b Calculate the daily average hours of sunshine for the whole year.

|  | Sydney | London |
|---|---|---|
| Mode |  |  |
| Mean |  |  |

c Which of the averages was more difficult to estimate? Why?
_____
_____

d Is the mode closer to the median for Sydney or London? _____

e The colder months in London are October to March. What is the difference between the daily average hours of sunshine in the colder and warmer months of the year? _____

f The warmer months in Sydney are October to March. What is the difference between the daily average hours of sunshine in the colder and warmer months of the year? _____

# Extended practice

Is it best to use mode, median or mean to describe averages?

**1** Interpreting an average from a set of data can be useful, but it can also be misleading. Sometimes people interpret data in a way that suits them.

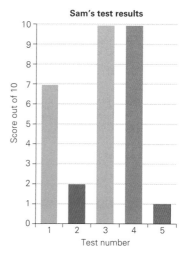

This graph shows Sam's spelling test scores over five weeks. Sam uses the mode to describe the average score. He tells his family, "My average score in spelling tests is 10 out of 10."

a Is Sam correct in saying that his average score is 10 out of 10?

_____

b Why is 10 out of 10 not a true reflection of Sam's average score?

_____

c What is Sam's average score as a median average? _____

d What is Sam's average score as a mean average? _____

---

**2** Here are Sam's scores out of 20 in the next 10 tests.

19, 20, 19, 19, 20, 20, 1, 19, 19, 19

a What is the range?       b What is the mode average?

_____                                      _____

c What is the median average?     d What is the mean average?

_____                                      _____

e Which average would you use to best describe Sam's level in spelling? Give a reason for your answer.

_____

---

**3** Alex recorded the temperature at midday for one week in summer. Complete the table so that the mean average works out to be 29 °C.

| Day | Temperature |
| --- | --- |
| Sunday | 28 °C |
| Monday | 29 °C |
| Tuesday |  |
| Wednesday | 24 °C |
| Thursday |  |
| Friday |  |
| Saturday | 27 °C |

# UNIT 10: TOPIC 1
# Describing probabilities

What is the likelihood of this 10-section spinner landing on blue?
You can describe the chance in different ways.

**In words:** It is unlikely.

**As a fraction:** There is a $\frac{1}{10}$ (or 1 out of 10) chance.

**As a percentage:** There is a 10% chance.

**As a decimal:** There is a 0.1 chance.

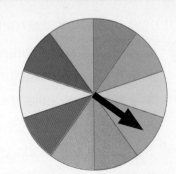

## Guided practice

*I'm certain you'll get all these right!*

**1** Using the probability words *certain, highly likely, likely, even chance, unlikely, highly unlikely* and *impossible*, describe the chance of the following things happening. (Try to use only one of each.)

    a    The next baby born will be a boy. _____

    b    You will receive a card for your birthday. _____

    c    You will fly to the moon 10 minutes from now. _____

    d    Somebody will smile in the next 10 minutes. _____

    e    New Year's Day will be on 1 January next year? _____

    f    You will appear on TV next month. _____

    g    You will hear a dog bark in the next 10 minutes. _____

This spinner is not divided equally.

**2** There is half a chance that spinner will land on red. Which fraction describes the probability of it landing on yellow? _____

**3** There is a 10% chance that the spinner will land on blue. Which percentage describes the probability of it landing on red? _____

**4** The probability as a decimal of the spinner landing on yellow is 0.1. Which decimal describes the probability of it landing on white? _____

# Independent practice

**1** A TV weather presenter says that there is not much chance of rain tomorrow. Which percentage best describes the probability? Underline one.

100%    0%    15%    50%    75%

*Try to choose appropriate probability descriptions.*

**2** Using a fraction for one, a percentage for another and a decimal for another, describe the probability of this spinner landing on:

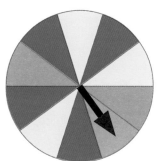

a green _____

b red _____

c blue _____

**3** What is the probability of the spinner in question 2 **not** landing on green? ☐

**4** Describe something that has the following chance of happening:

a 0.9 of a chance. _____

b 5% chance. _____

c 1 out of 2 chance. _____

d 0.25 of a chance. _____

e 100% chance. _____

**5** Colour this spinner so that the following probabilities are true.

- There is a 20% chance for yellow.
- There is a 3 out of 10 chance for blue.
- There is 0.2 of a chance for green.
- There is not much chance for red.
- There is a $\frac{1}{5}$ chance for white.

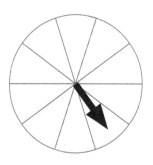

**6** Each jar contains 100 jelly beans. Write a value to show the probability of choosing (without looking) a white jelly bean from each jar. Choose from this list:

0.07    $\frac{4}{10}$    8%    $\frac{3}{4}$    0.8    $\frac{7}{10}$

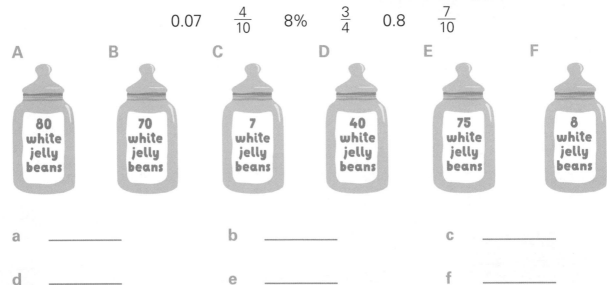

a _____    b _____    c _____

d _____    e _____    f _____

**7** Which of these does **not** show the chance of the spinner landing on blue? Circle one.

$\frac{1}{4}$    $\frac{4}{10}$    25%    0.25

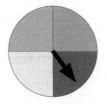

**8** Amy has to choose a bead without looking. Colour the beads so that she has:

- $\frac{1}{6}$ probability of choosing a red bead.
- $33\frac{1}{3}$% chance of choosing a yellow bead.
- 0.5 chance of choosing a blue bead.

**9** There is a mix of blue and yellow marbles in each bag of 100. Jack took 20 marbles from each bag without looking. Use Jack's results to predict the number of blue and yellow marbles in each bag of 100.

| Bag | After 20 have been taken out: | | My prediction after 100 have been taken out: | |
|---|---|---|---|---|
| A | Blue: 5 | Yellow: 15 | Blue: | Yellow: |
| B | Blue: 12 | Yellow: 8 | Blue: | Yellow: |
| C | Blue: 18 | Yellow: 2 | Blue: | Yellow: |
| D | Blue: 10 | Yellow: 10 | Blue: | Yellow: |

# Extended practice

**Tran's game: Part 1**

Tran plays a game of chance with some friends, in which a wheel spins and a ball lands on one of 37 numbers. Some players choose to guess what number the ball will land on. If they get it right, they win counters.

However, Tran wants to be the one who wins in the end. So he works out the likelihood of the ball landing on particular numbers. Zero is a green number and numbers 1 to 36 are either red or black.

**1**
  **a** The chance of the ball landing on a particular number is one out of: ☐

  **b** If 37 people each choose a different number and put a counter on their number on the table, how many counters does Tran have now? ☐

  **c** The rules say that Tran gives out 35:1 to someone choosing the correct number. So, the winner gets their counter back, plus 35 more counters. However, Tran does not lose any counters. Explain why.

  _____

  **d** If this continues for 1000 turns, Tran has done nothing except spin the wheel, but how many counters does he have now? ☐

**2** Tran realises that not many people would play if they only had a 1:37 chance of winning, so he thinks of other ways to get people to play. People can choose "red" or "black" numbers.

  **a** The chance of the ball landing on a red number is almost (but not quite) 1 out of 2. Explain why. _____

  _____

  **b** Imagine 18 out of the 37 people in question 1 choose red, 18 choose black and one person puts their counter on the green zero. If the ball lands on a black number, how many people lose? ☐

  **c** If the ball lands on black, 18 people win. However, Tran still does not lose any counters. Explain why.

  _____

  **d** If this continues for 10 000 turns, Tran has done nothing except spin the wheel, but how many counters does he win? ☐

# UNIT 10: TOPIC 2
## Conducting chance experiments and analysing outcomes

You will need some dice for these activities.

Working out number values for the chance of something occurring does not necessarily mean that it will happen that way. There is only a 1 out of 6 chance of the dice landing on a 6.

### Guided practice

*Is it harder to roll a six than a one?*

**1**  **a**  What is the chance of **not** rolling a six on one dice? _____

   **b**  Although it is unlikely, explain how somebody could get a six on the first roll of the dice. _____

---

**2**  **a**  You are going to roll a dice. Predict the number of rolls it will take until you roll a six. _____

   **b**  Roll a dice until you get a six. How many rolls did it take? _____

   **c**  What was the difference between your prediction and reality? Try to explain it.

   _____

   _____

---

**3**  **a**  Complete the table to show the number of times the dice should land on each number if it is rolled 36 times.

   **b**  Roll the dice 36 times and record the results.

   **c**  Write a sentence or two commenting on the results of the experiment.

| Dice lands on: | Number of times it will land like that | |
|---|---|---|
| | Probability | Actual number |
| 1 | | |
| 2 | | |
| 3 | | |
| 4 | | |
| 5 | | |
| 6 | | |

_____

_____

# Independent practice

**1** If you roll two dice, there is only one way for the dice to land to give the highest possible total of 12.

a   What is the lowest possible total? _____

b   How many ways can the dice land to give the lowest total? _____

**2** If you play a game with two 6-sided dice and you need to roll 11 to win the game, there are two ways the dice can land.

What are all the possible totals for two dice? Complete the table.

| Total of two dice | Ways the dice can land | Total number of ways |
|---|---|---|
| 12 | 6 + 6 | 1 |
| 11 | 6 + 5, 5 + 6 | 2 |
| 10 | | |
| 9 | | |
| 8 | | |
| 7 | | |
| 6 | | |
| 5 | | |
| 4 | | |
| 3 | | |
| 2 | | |

*There are 36 different ways the dice can land.*

**3** Which total has the best chance of being rolled? _____

**4** There is a 1 out of 36 chance of getting a total of 12 with two dice. What are the chances of a total of:

a  11: _____   b  10: _____   c  9: _____

d  8: _____    e  7: _____    f  6: _____

g  5: _____    h  4: _____    i  3: _____

j  2: _____    k  1: _____

**5** For this activity you will need two dice. You will be rolling the dice 72 times. Write the probable number of times that the dice should make each total. Carry out the experiment and write the actual totals.

| Total of two dice | Probable number of times out of 72 | Actual number of times out of 72 |
|---|---|---|
| 12 | 2 | |
| 11 | | |
| 10 | | |
| 9 | | |
| 8 | | |
| 7 | | |
| 6 | | |
| 5 | | |
| 4 | | |
| 3 | | |
| 2 | | |

**6** Colour each spinner according to these rules. There must be:

a  < 25% chance of it landing on yellow.

b  > 50% but < 75% chance of it landing on blue.

c  > 25% but < 50% chance of it landing on red.

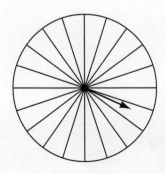
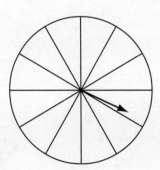

# Extended practice

Play Tran's game: Part 2

**You will need:**
- A 7-sided spinner (Trace it and glue onto card)
- Seven players with 10 counters each
- A "banker" with a bank of 50 counters
- Seven cards numbered 0–6 for each player.

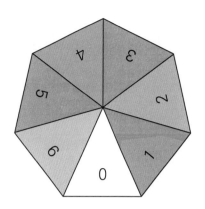

**1** The "banker" needs to find out the chances of someone winning.

    **a** What is the chance of the spinner landing on any particular number? _____

    **b** The person guessing the correct number receives six counters. If seven people choose a different number each and the spinner lands on six, how much does the "banker" put in the bank? _____

## How to play the game

**2** The "banker" writes, "My starting balance is 50 counters" on a sheet of paper. Each player draws a Win–Lose table with 13 rows and 5 columns similar to this:

| My Win–Lose Table ||||||
| --- | --- | --- | --- | --- |
| Turn | Guess | Win | Lose | Balance |
| Starting balance: | | | | 10 counters |
| 1st turn | 1 counter | | | |
| 2nd turn | 1 counter | | | |
| etc. | | | | |

**3** Each player guesses the number the spinner will land on by placing a card with that number on it in front of themselves.

    **a** Each player puts one counter in the middle and writes this in the "guess" column.

    **b** Somebody spins the spinner.

    **c** The person with the winning number gets six counters. The "banker" gets whatever is left over.

    **d** Players complete the row on their Win–Lose table. The "banker" writes their own new balance.

    **e** Repeat Steps 2 to 7 until the end of the tenth turn.

**4** Complete the sentences.

- My final balance was _____ counters.
- The balance for _____ players had decreased by the end of the game.
- The balance for the "banker" increased/decreased (underline one).

# GLOSSARY

**acute angle** An angle that is smaller than a right angle or 90 degrees.

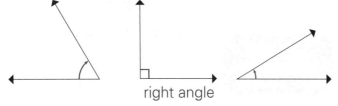
right angle

**addition** The joining or adding of two numbers together to find the total. Also known as *adding*, *plus* and *sum*. See also *vertical addition*.

★★★ + ★★ = ★★★★★
3 and 2 is 5

**algorithm** A process or formula used to solve a problem in mathematics.

Examples:
horizontal algorithms
24 + 13 = 37

vertical algorithms

|   | T | O |
|---|---|---|
|   | 2 | 4 |
| + | 1 | 3 |
|   | 3 | 7 |

**analogue time** Time shown on a clock or watch face with numbers and hands to indicate the hours and minutes.

**angle** The space between two lines or surfaces at the point where they meet, usually measured in degrees.

75-degree angle

**anticlockwise** Moving in the opposite direction to the hands of a clock.

**area** The size of an object's surface.

Example: It takes 12 tiles to cover this poster.

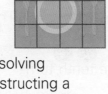

**area model** A visual way of solving multiplication problems by constructing a rectangle with the same dimensions as the numbers you are multiplying and breaking the problem down by place value.

6 × 10 = 60
6 × 8 = 48
so
6 × 18 = 108

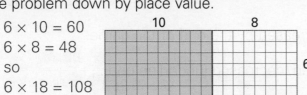

**array** An arrangement of items into even columns and rows to make them easier to count.

**balance scale** Equipment that balances items of equal mass; used to compare the mass of different items. Also called *pan balance* or *equal arm balance*.

**bar graph** A way of representing data using bars or columns to show the values of each variable.

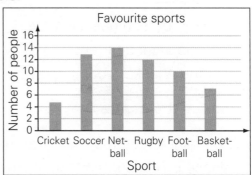

**base** The bottom edge of a 2D shape or the bottom face of a 3D shape.

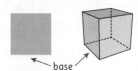

**capacity** The amount that a container can hold.

Example: The jug has a capacity of 4 cups.

**Cartesian plane** A grid system with numbered horizontal and vertical axes that allow for exact locations to be described and found.

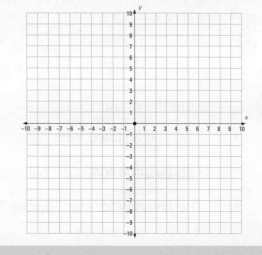

**categorical variables** The different groups that objects or data can be sorted into based on common features.

Example: Within the category of ice-cream flavours, variables include:

vanilla      chocolate      strawberry

**centimetre** or **cm** A unit for measuring the length of smaller items.

Example: Length is 80 cm.

**circle graph** A circular graph divided into sections that look like portions of a pie.

**circumference** The distance around the outside of a circle.

**clockwise** Moving in the same direction as the hands of a clock.

**common denominator** Denominators that are the same. To find a common denominator, you need to identify a multiple that two or more denominators share.

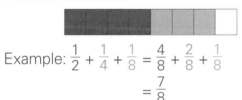

Example: $\frac{1}{2} + \frac{1}{4} + \frac{1}{8} = \frac{4}{8} + \frac{2}{8} + \frac{1}{8}$
$= \frac{7}{8}$

**compensation strategy** A way of solving a problem that involves rounding a number to make it easier to work with, and then paying back or "compensating" the same amount.

Example: 24 + 99 = 24 + 100 − 1 = 123

**composite number** A number that has more than two factors, that is, a number that is not a prime number.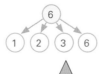

**cone** A 3D shape with a circular base that tapers to a point.

**coordinates** A combination of numbers or numbers and letters that show location on a grid map.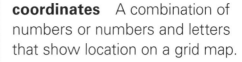

**corner** The point where two edges of a shape or object meet. Also known as a *vertex*.

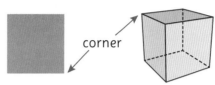

**cross-section** The surface or shape that results from making a straight cut through a 3D shape.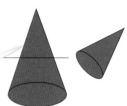

**cube** A rectangular prism where all six faces are squares of equal size.

**cubic centimetre** or **cm³** A unit for measuring the volume of smaller objects.

Example: This cube is exactly 1 cm long, 1 cm wide and 1 cm deep.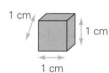

**cylinder** A 3D shape with two parallel circular bases and one curved surface.

**data** Information gathered through methods such as questioning, surveys or observation.

**decimal fraction** A way of writing a number that separates any whole numbers from fractional parts expressed as tenths, hundredths, thousandths and so on.

Example: 1.9 is the same as 1 whole and 9 parts out of 10 or $1\frac{9}{10}$.

**degrees Celsius** A unit used to measure the temperature against the Celsius scale where 0°C is the freezing point and 100°C is the boiling point.

**denominator** The bottom number in a fraction, which shows how many pieces the whole or group has been divided into.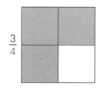

**diameter** A straight line from one side of a circle to the other, passing through the centre point.

**digital time** Time shown on a clock or watch face with numbers only to indicate the hours and minutes.

**division/dividing** The process of sharing a number or group into equal parts, with or without remainders.

**dot plot** A way of representing pieces of data using dots along a line labelled with variables.

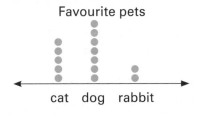

**double/doubles** Adding two identical numbers or multiplying a number by 2.
   Example:   $2 + 2 = 4$     $4 \times 2 = 8$

**duration** How long something lasts.
   Example: Most movies have a duration of about 2 hours.

**edge** The side of a shape or the line where two faces of an object meet.

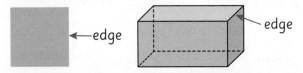

**equal** Having the same number or value.

   Example:   Equal size      Equal numbers

**equation** A written mathematical problem where both sides are equal.
   Example:     $4 + 5 = 6 + 3$

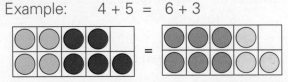

**equilateral triangle** A triangle with three sides and angles the same size.

**equivalent fractions** Different fractions that represent the same size in relation to a whole or group.

 $\frac{1}{2}$    $\frac{2}{4}$    $\frac{3}{6}$    $\frac{4}{8}$

**estimate** A thinking guess.

**even number** A number that can be divided equally into 2.
   Example: 4 and 8 are even numbers

**face** The flat surface of a 3D shape.

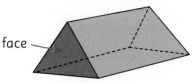

**factor** A whole number that will divide evenly into another number.
   Example: The factors of 10 are  1 and 10
                                    2 and 5

**financial plan** A plan that helps you to organise or manage your money.

**flip** To turn a shape over horizontally or vertically. Also known as *reflection*.

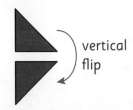 

**fraction** An equal part of a whole or group.
   Example: One out of two parts or $\frac{1}{2}$ is shaded.

**grams or *g*** A unit for measuring the mass of smaller items.

1000 g is 1 kg

**graph** A visual way to represent data or information.

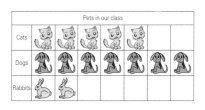 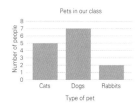

**GST or Goods and Services Tax** A tax, such as 10%, that applies to most goods and services bought in many countries.

Example: Cost + GST (10%) = Amount you pay
$10 + $0.10 = $10.10

**hexagon** A 2D shape with six sides.

**horizontal** Parallel with the horizon or going straight across.

**improper fraction** A fraction where the numerator is greater than the denominator, such as $\frac{3}{2}$.

**integer** A whole number. Integers can be positive or negative.

**inverse operations** Operations that are the opposite or reverse of each other. Addition and subtraction are inverse operations.

Example: 6 + 7 = 13 can be reversed with 13 − 7 = 6

**invoice** A written list of goods and services provided, including their cost and any GST.

| Priya's Pet Store | | | |
|---|---|---|---|
| **Tax Invoice** | | | |
| Item | Quantity | Unit price | Cost |
| Siamese cat | 1 | $500 | $500.00 |
| Cat food | 20 | $1.50 | $30.00 |
| Total price of goods | | | $530.00 |
| GST (10%) | | | $53.00 |
| Total | | | $583.00 |

**isosceles triangle** A triangle with two sides and two angles of the same size.

**jump strategy** A way to solve number problems that uses place value to "jump" along a number line by hundreds, tens and ones.

Example: 16 + 22 = 38

**kilograms or kg** A unit for measuring the mass of larger items.

**kilometres or km** A unit for measuring long distances or lengths.

**kite** A four-sided shape where two pairs of adjacent sides are the same length.

**legend** A key that tells you what the symbols on a map mean.

**length** The longest dimension of a shape or object.

**line graph** A type of graph that joins plotted data with a line.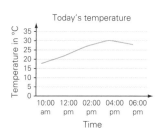

**litres or L**  A unit for measuring the capacity of larger containers.

Example: The capacity of this bucket is 8 litres.

**mass**  How heavy an object is.

Example:   4.5 kilograms   4.5 grams

**mean**  The total of a set of numbers divided by however many numbers there are in the set.

Example: 5, 3, 6, 2, 4 – the mean is 20 ÷ 5 = 4

**median**  The number in the middle of an ordered set of numbers.

Example: 3, 4, 5, 6, 7 – the median is 5

**metre or m**  A unit for measuring the length of larger objects.

**milligram or mg**  A unit for measuring the mass of lighter items or to use when accuracy of measurements is important.

700 mg

**millilitre or mL**  A unit for measuring the capacity of smaller containers.

1000 mL is 1 litre

**millimetre or mm**  A unit for measuring the length of very small items or to use when accuracy of measurements is important.

There are 10 mm in 1 cm.

**mixed number**  A number that contains both a whole number and a fraction.

Example: $2\frac{3}{4}$

**mode**  The number that occurs most often in a set of numbers.

Example: 2, 3, 2, 5, 2 – the mode is 2

**multiple**  The result of multiplying a particular whole number by another whole number.

Example: 10, 15, 20 and 100 are all multiples of 5.

**near doubles**  A way to add two nearly identical numbers by using known doubles facts.

Example: 4 + 5 = 4 + 4 + 1 = 9

**net**  A flat shape that when folded up makes a 3D shape.

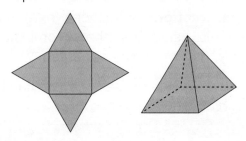

**number line**  A line on which numbers can be placed to show their order in our number system or to help with calculations.

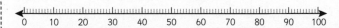

**number sentence**  A way to record calculations using numbers and mathematical symbols.

Example: 23 + 7 = 30

**numeral**  A figure or symbol used to represent a number.

Examples: 1 – one   2 – two   3 – three

**numerator**  The top number in a fraction, which shows how many pieces you are dealing with.

**obtuse angle** An angle that is larger than a right angle or 90 degrees, but smaller than 180 degrees.

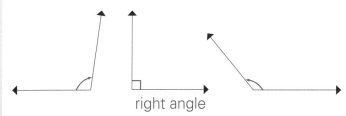

**octagon** A 2D shape with eight sides.

**odd number** A number that cannot be divided equally into 2.

Example: 5 and 9 are odd numbers.

**operation** A mathematical process. The four basic operations are addition, subtraction, multiplication and division.

**origin** The point on a Cartesian plane where the x-axis and y-axis intersect.

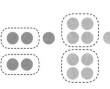

**outcome** The result of a chance experiment.

Example: The possible outcomes if you roll a dice are 1, 2, 3, 4, 5 or 6.

**parallel lines** Straight lines that are the same distance apart and so will never cross.

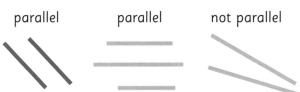

**parallelogram** A four-sided shape where each pair of opposite sides is parallel.

**pattern** A repeating design or sequence of numbers.

Example:
Shape pattern
Number pattern 2, 4, 6, 8, 10, 12

**pentagon** A 2D shape with five sides.

**per cent or %** A fraction out of 100.

Example: $\frac{62}{100}$ or 62 out of 100

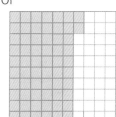

is also 62%.

**perimeter** The distance around the outside of a shape or area.

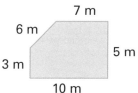

Example: Perimeter = 7 m + 5 m + 10 m + 3 m + 6 m = 31 m

**pictograph** A way of representing data using pictures so that it is easy to understand.

Example: Favourite juices in our class

**place value** The value of a digit depending on its place in a number.

| M | H Th | T Th | Th | H | T | O |
|---|------|------|----|----|---|---|
|   |      |      | 2  | 7 | 4 | 8 |
|   |      | 2    | 7  | 4 | 8 | 6 |
|   | 2    | 7    | 4  | 8 | 6 | 3 |
| 2 | 7    | 4    | 8  | 6 | 3 | 1 |

**polygon** A closed 2D shape with three or more straight sides.

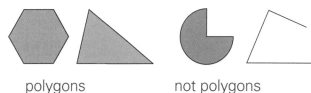

polygons          not polygons

**polyhedron (plural polyhedra)** A 3D shape with flat faces.

polyhedra   not polyhedra

**power of** The number of times a particular number is multiplied by itself.

Example: $4^3$ is 4 to the power of 3 or $4 \times 4 \times 4$.

**prime number** A number that has just two factors – 1 and itself. The first four prime numbers are 2, 3, 5 and 7.

**prism** A 3D shape with parallel bases of the same shape and rectangular side faces.

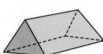 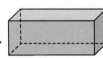 

triangular prism   rectangular prism   hexagonal prism

**probability** The chance or likelihood of a particular event or outcome occurring.

 Example: There is a 1 in 8 chance this spinner will land on red.

**protractor** An instrument used to measure the size of angles in degrees.

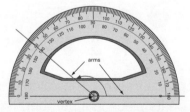

**pyramid** A 3D shape with a 2D shape as a base and triangular faces meeting at a point.

square pyramid   hexagonal pyramid

**quadrant** A quarter of a circle or one of the four quarters on a Cartesian plane.

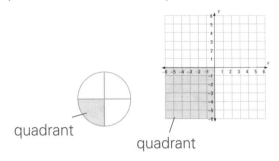

quadrant   quadrant

**quadrilateral** Any 2D shape with four sides.

**radius** The distance from the centre of a circle to its circumference or edge.

**range** The difference between the highest and lowest in a set of numbers.

Example: 5, 3, 6, 2, 4 – the range is 6 – 2 = 4

**reflect** To turn a shape over horizontally or vertically. Also known as *flipping*.

vertical reflection   horizontal reflection

**reflex angle** An angle that is between 180 and 360 degrees in size.

**remainder** An amount left over after dividing one number by another.

Example: 11 ÷ 5 = 2 r1

**rhombus** A 2D shape with four sides, all of the same length and opposite sides parallel.

**right angle** An angle of exactly 90 degrees.

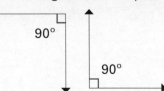

**right-angled triangle**  A triangle where one angle is exactly 90 degrees.

 **rotate**  Turn around a point.

**rotational symmetry**  A shape has rotational symmetry if it fits into its own outline at least once while being turned around a fixed centre point.

1st position    Back to the start

2nd position

**round/rounding**  To change a number to another number that is close to it to make it easier to work with.

229 can be

rounded up to the nearest 10   OR   rounded down to the nearest 100

↑ 230          ↓ 200

**scale**  A way to represent large areas on maps by using ratios of smaller to larger measurements.

Example: 1 cm = 5 m

**scalene triangle**  A triangle where no sides are the same length and no angles are equal.

**sector**  A section of a circle bounded by two radius lines and an arc.

radius lines — sector — arc

**semi-circle**  Half a circle, bounded by an arc and a diameter line.

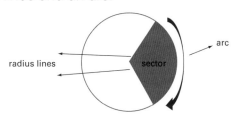
semi-circle
arc
diameter line

**skip counting**  Counting forwards or backwards by the same number each time.

Examples:
Skip counting by fives: 5, 10, 15, 20, 25, 30
Skip counting by twos: 1, 3, 5, 7, 9, 11, 13

**slide**  To move a shape to a new position without flipping or turning it. Also known as *translate*.

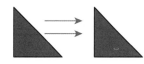

 **sphere**  A 3D shape that is perfectly round.

**split strategy**  A way to solve number problems that involves splitting numbers up using place value to make them easier to work with.

Example: 21 + 14 =
20 + 10 + 1 + 4 = 35

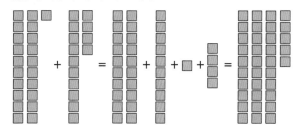

**square centimetre or $cm^2$**  A unit for measuring the area of smaller objects. It is exactly 1 cm long and 1 cm wide.

**square metre or $m^2$**  A unit for measuring the area of larger spaces. It is exactly 1 m long and 1 m wide.

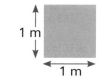

**square number**  The result of a number being multiplied by itself. The product can be represented as a square array.

Example: 3 × 3 or $3^2$ = 9

**straight angle**  An angle that is exactly 180 degrees in size.

180°

**strategy** A way to solve a problem. In mathematics, you can often use more than one strategy to get the right answer.

Example: 32 + 27 = 59

Jump strategy

Split strategy

30 + 2 + 20 + 7 = 30 + 20 + 2 + 7 = 59

**subtraction** The taking away of one number from another number. Also known as *subtracting, take away, difference between* and *minus*. See also *vertical subtraction*.

Example: 5 take away 2 is 3

**survey** A way of collecting data or information by asking questions.

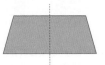

**symmetry** A shape or pattern has symmetry when one side is a mirror image of the other.

**table** A way to organise information that uses columns and rows.

| Flavour | Number of people |
|---|---|
| Chocolate | 12 |
| Vanilla | 7 |
| Strawberry | 8 |

**tally marks** A way of keeping count that uses single lines with every fifth line crossed to make a group.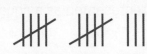

**term** A number in a series or pattern.

Example: The sixth term in this pattern is 18.

| 3 | 6 | 9 | 12 | 15 | 18 | 21 | 24 |

**tessellation** A pattern formed by shapes that fit together without any gaps.

**thermometer** An instrument for measuring temperature.

**three-dimensional or 3D** A shape that has three dimensions – length, width and depth. 3D shapes are not flat.

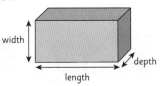

**time line** A visual representation of a period of time with significant events marked in.

**translate** To move a shape to a new position without flipping or turning it. Also known as *slide*.

**trapezium** A 2D shape with four sides and only one set of parallel lines.

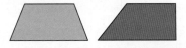

**triangular number** A number that can be organised into a triangular shape. The first four are:

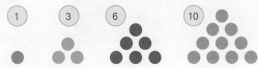

**two-dimensional or 2D** A flat shape that has two dimensions – length and width.

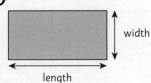

**turn** Rotate around a point.

**unequal** Not having the same size or value.
Example: Unequal size   Unequal numbers
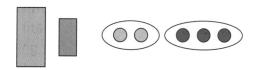

**value** How much something is worth.
Example:
This coin is worth 5c.   This coin is worth $1.

**vertex (plural vertices)** The point where two edges of a shape or object meet. Also known as a *corner*.
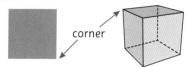

**vertical** At a right angle to the horizon or straight up and down.
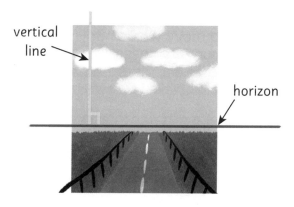

**vertical addition** A way of recording addition so that the place-value columns are lined up vertically to make calculation easier.

| T | O |
|---|---|
| 3 | 6 |
| + 2 | 1 |
| 5 | 7 |

**vertical subtraction** A way of recording subtraction so that the place-value columns are lined up vertically to make calculation easier.

| T | O |
|---|---|
| 5 | 7 |
| − 2 | 1 |
| 3 | 6 |

**volume** How much space an object takes up.
 Example: This object has a volume of 4 cubes.

**whole** All of an item or group.
Example: A whole shape   A whole group

**width** The shortest dimension of a shape or object. Also known as *breadth*.

**x-axis** The horizontal reference line showing coordinates or values on a graph or map.
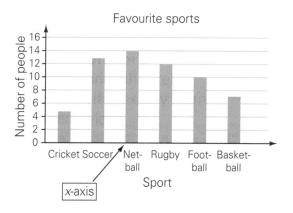

**y-axis** The vertical reference line showing coordinates or values on a graph or map.
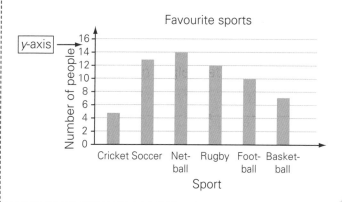

# ANSWERS

## UNIT 1: Topic 1

### Guided practice

1

| M | Hth | Tth | Th | H | T | O | Number |
|---|---|---|---|---|---|---|---|
| 5 | 0 | 0 | 0 | 0 | 0 | 0 | 5 000 000 |
| | 3 | 0 | 0 | 0 | 0 | 0 | 300 000 |
| | | 6 | 0 | 0 | 0 | 0 | 60 000 |
| | | | 7 | 0 | 0 | 0 | 7000 |
| | | | | 9 | 0 | 0 | 900 |
| | | | | | 1 | 0 | 10 |
| | | | | | | 8 | 8 |

2 a 51 604
b 200 026
c 12 010

### Independent practice

1 a 60 000 b 300 000
c 6000 d 1 000 000
e 80 000 000 f 50 000 000

2 a sixty thousand
b three hundred thousand
c six thousand d one million
e eighty million f fifty million

3 a 80 487 000 b 10 362 059
c 114 760 209 d 1 400 593 001

4 a As Student Book
b 200 000 + 10 000 + 4000 + 800 + 60 + 7
c 2 000 000 + 500 000 + 60 000 + 7000 + 300 + 20 + 1
d 5 000 000 + 600 000 + 70 000 + 3000 + 200 + 7
e 50 000 000 + 7 000 000 + 300 000 + 10 000 + 9000 + 200 + 40
f 400 000 000 + 7 000 000 + 500 000 + 8000 + 4

5 a 9 754 321 b 5 123 479
c 9 543 217 d 2 314 579

6 a 6 142 793: six million, one hundred and forty-two thousand, seven hundred and ninety-three
b 280 526 306: two hundred and eighty million, five hundred and twenty-six thousand, three hundred and six

### Extended practice

1 a + 100 b + 40 000
c − 20 000 d + 1

2 a $340 000 b $705 000
c $825 000 d $1 250 000

3 Answers may vary. Look for students who justify their answers appropriately. Possible answers:
a The digit 5. It means 500 000. $500 000 is a lot of money!
b The digit 2. It means 2 whole ones. I wouldn't want to write out my times tables any more than that!
c The digit 1. It means 10. I really like them but too many might make me ill!

## UNIT 1: Topic 2

### Guided practice

1 $3 \times 3 = 3^2$, $3^2 = 9$, $4 \times 4 = 4^2$, $4^2 = 16$, $5 \times 5 = 5^2$, $5^2 = 25$, $6 \times 6 = 6^2$, $6^2 = 36$

2 1 + 2 + 3 = 6, 1 + 2 + 3 + 4 = 10

### Independent practice

1 (Shading may vary.) $7 \times 7 = 7^2$, $8 \times 8 = 8^2$, $9 \times 9 = 9^2$, $10 \times 10 = 10^2$, $7^2 = 49$, $8^2 = 64$, $9^2 = 81$, $10^2 = 100$

2 a 121
b Teacher to check, e.g. The digits alternate between odd and even.
c 10 000

3 Teacher to check artwork.
15: 1 + 2 + 3 + 4 + 5 = 15,
21: 1 + 2 + 3 + 4 + 5 + 6 = 21,
28: 1 + 2 + 3 + 4 + 5 + 6 + 7 = 28,
36: 1 + 2 + 3 + 4 + 5 + 6 + 7 + 8 = 36,
45: 1 + 2 + 3 + 4 + 5 + 6 + 7 + 8 + 9 = 45,
55: 1 + 2 + 3 + 4 + 5 + 6 + 7 + 8 + 9 + 10 = 55

4 a 66 b 36
c Teacher to check, e.g. The numbers alternate between two odd numbers, then two even numbers.

### Extended practice

1

| Square number | Multiplication fact | Addition fact |
|---|---|---|
| $1^2 = 1$ | 1 × 1 = 1 | 1 |
| $2^2 = 4$ | 2 × 2 = 4 | 1 + 3 = 4 |
| $3^2 = 9$ | 3 × 3 = 9 | 1 + 3 + 5 = 9 |
| $4^2 = 16$ | 4 × 4 = 16 | 1 + 3 + 5 + 7 = 16 |
| $5^2 = 25$ | 5 × 5 = 25 | 1 + 3 + 5 + 7 + 9 = 25 |
| $6^2 = 36$ | 6 × 6 = 36 | 1 + 3 + 5 + 7 + 9 + 11 = 36 |
| $7^2 = 49$ | 7 × 7 = 49 | 1 + 3 + 5 + 7 + 9 + 11 + 13 = 49 |
| $8^2 = 64$ | 8 × 8 = 64 | 1 + 3 + 5 + 7 + 9 + 11 + 13 + 15 = 64 |
| $9^2 = 81$ | 9 × 9 = 81 | 1 + 3 + 5 + 7 + 9 + 11 + 13 + 15 + 17 = 81 |
| $10^2 = 100$ | 10 × 10 = 100 | 1 + 3 + 5 + 7 + 9 + 11 + 13 + 15 + 17 + 19 = 100 |

2 a Teacher to check, e.g. To get to the next square number, you add the next odd number.
b $11^2 = 121$, 11 × 11 = 121, 1 + 3 + 5 + 7 + 9 + 11 + 13 + 15 + 17 + 19 + 21 = 121
c 23

3 a 15 b 1, 5, 12, 22, 35
c Teacher to check, e.g. To get to the second pentagonal number you add 4 to the first one (1 + 4 = 5). For the next one you add 3 more than that (5 + 7 = 12). Each time, you add 3 more than last time.
d

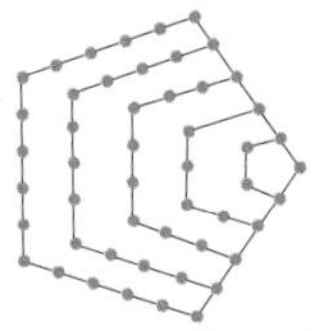

## UNIT 1: Topic 3

### Guided practice

1

| Number | Factors (numbers it can be divided by) | How many factors? | Prime or composite? Prime | Composite |
|---|---|---|---|---|
| 1 | 1 | 1 | neither | |
| 2 | 1 and 2 | 2 | ✔ | |
| 3 | 1 & 3 | 2 | ✔ | |
| 4 | 1, 2 & 4 | 3 | | ✔ |
| 5 | 1 & 5 | 2 | ✔ | |
| 6 | 1, 6, 2, 3 | 4 | | ✔ |
| 7 | 1 & 7 | 2 | ✔ | |
| 8 | 1, 8, 2, 4 | 4 | | ✔ |
| 9 | 1, 9, 3 | 3 | | ✔ |
| 10 | 1, 10, 2, 5 | 4 | | ✔ |
| 11 | 1 & 11 | 2 | ✔ | |
| 12 | 1, 12, 2, 6, 3, 4 | 6 | | ✔ |
| 13 | 1 & 13 | 2 | ✔ | |
| 14 | 1, 14, 2, 7 | 4 | | ✔ |
| 15 | 1, 15, 3, 5 | 4 | | ✔ |
| 16 | 1, 16, 2, 8, 4 | 5 | | ✔ |
| 17 | 1 & 17 | 2 | ✔ | |
| 18 | 1, 18, 2, 9, 3, 6 | 6 | | ✔ |
| 19 | 1 & 19 | 2 | ✔ | |
| 20 | 1, 20, 2, 10, 4, 5 | 6 | | ✔ |

2 a 2, 3, 5, 7, 11, 13, 17, 19
b Teacher to check, e.g. 2 is the only even prime number.

### Independent practice

1

| | | | | | | | | | |
|---|---|---|---|---|---|---|---|---|---|
| 1 | **2** | **3** | 4 | **5** | 6 | **7** | 8 | 9 | 10 |
| **11** | 12 | **13** | 14 | 15 | 16 | **17** | 18 | **19** | 20 |
| 21 | 22 | **23** | 24 | 25 | 26 | 27 | 28 | **29** | 30 |
| **31** | 32 | 33 | 34 | 35 | 36 | **37** | 38 | 39 | 40 |
| **41** | 42 | **43** | 44 | 45 | 46 | **47** | 48 | 49 | 50 |
| 51 | 52 | **53** | 54 | 55 | 56 | 57 | 58 | **59** | 60 |
| **61** | 62 | 63 | 64 | 65 | 66 | **67** | 68 | 69 | 70 |
| **71** | 72 | **73** | 74 | 75 | 76 | 77 | 78 | **79** | 80 |
| 81 | 82 | **83** | 84 | 85 | 86 | 87 | 88 | **89** | 90 |
| 91 | 92 | 93 | 94 | 95 | 96 | **97** | 98 | 99 | 100 |

2 a 97 b False
c False (49 even composite numbers, 25 odd composite numbers).

**3**

a The prime factors of 10 are 2 and 5. So 2 × 5 = 10

b The prime factors of 9 are 3 and 3. So 3 × 3 = 9

c The prime factors of 15 are 3 and 5. So 3 × 5 = 15

d The prime factors of 21 are 3 and 7. So 3 × 7 = 21

e The prime factors of 35 are 5 and 7. So 5 × 7 = 35

f The prime factors of 39 are 3 and 13. So 3 × 13 = 39

g The prime factors of 26 are 2 and 13. So 2 × 13 = 26

h The prime factors of 33 are 3 and 11. So 3 × 11 = 33

i The prime factors of 34 are 2 and 17. So 2 × 17 = 34

**4**

a The prime factors of 14 are 2 and 7. So 2 × 7 = 14

b The prime factors of 55 are 5 and 11. So 5 × 11 = 55

c The prime factors of 49 are 7 and 7. So 7 × 7 = 49

## Extended practice

**1**

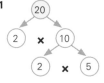

a  20 = 2 × 2 × 5
   20 = $2^2$ × 5

b  18 = 3 × 3 × 2
   18 = $3^2$ × 2

c  28 = 2 × 2 × 7
   28 = $2^2$ × 7

d  36 = 2 × 2 × 3 × 3
   36 = $2^2$ × $3^2$

**2** Diagrams may vary, e.g. the second row of c) could show 3 × 8. Teacher to check the resulting prime factors.

a  27 = 3 × 3 × 3
   27 = $3^3$

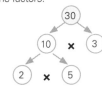
b  30 = 2 × 5 × 3

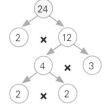
c  24 = 2 × 2 × 2 × 3
   24 = $2^3$ × 3

## UNIT 1: Topic 4

### Guided practice

**1**

| Problem | Using rounding it becomes | | | Now I need to | Answer |
|---|---|---|---|---|---|
| a | 317 + 199 | 317 + 200 = 517 | | take away 1 | 516 |
| b | 275 − 101 | 275 − 100 = 175 | | take away another 1 | 174 |
| c | 527 + 302 | 527 + | 300 = | 827 | add another 2 | 829 |
| d | 377 − 98 | 377 − | 100 = | 277 | add back 2 | 279 |
| e | 249 + 249 | 250 + 250 = 500 | | take away 2 | 498 |
| f | 938 − 206 | 938 − 200 = 738 | | take away another 6 | 732 |
| g | 1464 + 998 | 1464 + 1000 = 2464 | | take away 2 | 2462 |

**2**

| Problem | Expand the numbers | Join the partners | Answer |
|---|---|---|---|
| a | 370 + 520 | 300 + 70 + 500 + 20 | 300 + 500 + 70 + 20 | 890 |
| b | 2200 + 3600 | 2000 + 200 + 3000 + 600 | 2000 + 3000 + 200 + 600 | 5800 |
| c | 342 + 236 | 300 + 40 + 2 + 200 + 30 + 6 | 300 + 200 + 40 + 30 + 2 + 6 | 578 |
| d | 471 + 228 | 400 + 70 + 1 + 200 + 20 + 8 | 400 + 200 + 70 + 20 + 1 + 8 | 699 |
| e | 743 + 426 | 700 + 40 + 3 + 400 + 20 + 6 | 700 + 400 + 40 + 20 + 3 + 6 | 1169 |
| f | 865 + 734 | 800 + 60 + 5 + 700 + 30 + 4 | 800 + 700 + 60 + 30 + 5 + 4 | 1599 |
| g | 4270 + 3220 | 4000 + 200 + 70 + 3000 + 200 + 20 | 4000 + 3000 + 200 + 200 + 70 + 20 | 7490 |

### Independent practice

**1**  a 379     b 599     c 1298
      d 2284    e 3909    f 10 990

**2**  a 446     b 765     c 874
      d 545     e 768     f 1206

**3** Examples of strategies that might be used:
   a 898 (I said 650 + 250 = 900 then took away 2.)
   b 1054 (I took 200 from 1253 = 1053. Then I added 1 back.)
   c 3500 (I doubled 1500 then doubled 250.)
   d 14 168 (I took 400 away and then took another 10.)

**4** Examples of the way the numbers might be rounded:

| | Number fact | Rounded number | I rounded this number to the nearest… |
|---|---|---|---|
| a | Australia has 812 972 kilometres of roads. | 813 000 km | thousand |
| b | The Electricity Company of China employs 1 502 000 people. | 1 500 000 | hundred thousand |
| c | The Mexican soccer player, Blanco, earned $2 943 702 in 2009. | $3 000 000 | million |
| d | The fastest speed recorded at the Indianapolis 500 car race was 299.3 km/h. | 300 km/h | hundred |
| e | The fastest 100-metre sprint time for a woman is 10.49 seconds. | 10.5 | tenth |
| f | The US department store Walmart employs 2 100 000 people. | 2 000 000 | million |
| g | Each Australian eats an average of 17 L 600 mL of ice-cream a year. | 18 L | litre |
| h | The longest rail tunnel is in Switzerland. It is 57.1 km long. | 57 km | kilometre |
| i | The amount of money the movie *Avatar* made was $2 783 919 000. | $3 billion | billion |
| j | Foreign tourists spend $29 127 000 000 a year in Australia. | $29 billion or $30 billion | billion or ten billion |

**5**  Basic:     $22 490
      Deluxe:   $31 490
      Premium:  $38 959

# Extended practice

1 Examples of how the numbers might be rounded:

| | Problem | Round the numbers | Estimate the answer | Which is the likely answer? |
|---|---|---|---|---|
| e.g. | 109 897 + 50 157 | 110 000 + 50 000 | 160 000 | 261 054 or (161 054) |
| a | 5189 – 2995 | 5000 – 3000 | 2000 | 2194 or 3194 |
| b | 2958 + 6058 | 3000 + 6000 | 9000 | 9016 or 8016 |
| c | 8215 – 3108 | 8000 – 3000 | 5000 | 5907 or 5107 |
| d | 15 963 + 14 387 | 16 000 + 14 000 | 30 000 | 29 350 or 30 350 |
| e | 8954 – 3928 | 9000 – 4000 | 5000 | 5026 or 4026 |
| f | 4568 + 4489 | 4500 + 4500 | 9000 | 8057 or 9057 |
| g | 13 149 – 7908 | 13 000 – 8000 | 5000 | 6241 or 5241 |
| h | 124 963 + 98 358 | 125 000 + 100 000 | 225 000 | 223 321 or 213 321 |

2 Examples of ways of rounding:

| | Problem | Round the numbers | Estimate the answer | Calculator answer |
|---|---|---|---|---|
| e.g. | 6190 + 1880 | 6000 + 2000 | 8000 | 8070 |
| a | 4155 + 2896 | 4000 + 3000 | 7000 | 7051 |
| b | 9124 – 8123 | 9000 – 8000 | 1000 | 1001 |
| c | 24 065 + 5103 | 24 000 + 5000 | 29 000 | 29 168 |
| d | 19 753 – 10 338 | 20 000 – 10 000 | 10 000 | 9415 |
| e | 101 582 + 49 268 | 101 000 + 49 000 | 150 000 | 150 850 |
| f | 298 047 – 198 214 | 300 000 – 200 000 | 100 000 | 99 833 |
| g | 1 089 274 + 1 099 583 | 1 000 000 + 1 000 000 | 2 000 000 | 2 188 857 |
| h | 1 499 836 + 1 489 967 | 1 500 000 + 1 500 000 | 3 000 000 | 2 989 803 |

## Unit 1: Topic 5

### Guided practice

1 a 1234  b 2345  c 3456
  d 4567  e 5789  f 5678

### Independent practice

1 a 1111  b 2222  c 3333
  d 4444  e 5555  f 6666
  g 7777  h 8888  i 9999
  j 10 000  k 11 111

2 Look for the strategies used to solve the problem. One simple solution is to subtract 123 from 99 999 four times, making the addition 99 507 + 123 + 123 + 123 + 123.

3 a

| Country | Paved | Unpaved roads | Total |
|---|---|---|---|
| USA | 4 165 110 | 2 265 256 | 6 430 366 |
| India | 1 603 705 | 1 779 639 | 3 383 344 |
| China | 1 515 797 | 354 864 | 1 870 661 |
| France | 951 220 | 0 | 951 220 |
| Japan | 925 000 | 258 000 | 1 183 000 |
| Spain | 659 629 | 6 663 | 666 292 |
| Canada | 415 600 | 626 700 | 1 042 300 |
| Australia | 336 962 | 473 679 | 810 641 |
| Brazil | 96 353 | 1 655 515 | 1 751 868 |

  b India, Canada and Brazil
  c USA and China
  d China and Canada (981 564 km)

### Extended practice

1 947 344   2 562 996
3 The correct answer is 453 487. Teacher to check students' methods of confirming the correct answer.
4 335 358 000 000 km²

## UNIT 1: Topic 6

### Guided practice

1 a 229  b 326  c 2208  d 2119
2 a 589  b 199  c 2149
  d 1985  e 9988  f 8899

### Independent practice

1 a 54 321  b 65 432  c 76 543
  d 87 654  e 98 765  f 56 789
  g 45 678  h 34 567  i 23 456
  j 12 345

2 9 875 432 – 2 345 789 = 7 529 643

3 a 3268  b 12 619  c 22 656
  d 34 579  e 375 777  f 676 068
  g 749  h 3649  i 320 054
  j 65 622

### Extended practice

1 a 193 635  b 126 296  c 191 790
2 a Teacher to check; using any three digits will result in the answer 1089.
  b The answer is still 1089.

## Unit 1: Topic 7

### Guided practice

1

| × | 10 | 100 | 1000 | 10 000 |
|---|---|---|---|---|
| a | 29 | 290 | 2900 | 29 000 | 290 000 |
| b | 124 | 1240 | 12 400 | 124 000 | 1 240 000 |
| c | 638 | 6380 | 63 800 | 638 000 | 6 380 000 |
| d | $1.25 | $12.50 | $125 | $1250 | $12 500 |
| e | 750 | 7500 | 75 000 | 750 000 | 7 500 000 |

2

| | ÷ | 10 | Write the multiplication fact partner |
|---|---|---|---|
| a | 370 | 37 | 37 × 10 = 370 |
| b | 4700 | 470 | 470 × 10 = 4700 |
| c | 2000 | 200 | 200 × 10 = 2000 |
| d | $22.50 | $2.25 | $2.25 × 10 = $22.50 |
| e | 54 | 5.4 | 5.4 × 10 = 54 |

3

| | ÷ | 100 | Write the multiplication fact partner |
|---|---|---|---|
| a | 700 | 7 | 7 × 100 = 700 |
| b | $495 | $4.95 | $4.95 × 100 = $495 |
| c | 5000 | 50 | 50 × 100 = 5000 |
| d | 12 000 | 120 | 120 × 100 = 12 000 |
| e | 8750 | 87.5 | 87.5 × 100 = 8750 |

### Independent practice

1

| × | 10 | 20 [double] | 40 [double again] | 80 [double again] |
|---|---|---|---|---|
| a | 12 | 120 | 240 | 480 | 960 |
| b | 15 | 150 | 300 | 600 | 1200 |
| c | 22 | 220 | 440 | 880 | 1760 |
| d | 25 | 250 | 500 | 1000 | 2000 |
| e | 50 | 500 | 1000 | 2000 | 4000 |

2

| ÷ | 10 | 20 [halve it] | 40 [halve again] | 80 [halve again] |
|---|---|---|---|---|
| a | 400 | 40 | 20 | 10 | 5 |
| b | 2000 | 200 | 100 | 50 | 25 |
| c | 480 | 48 | 24 | 12 | 6 |
| d | 10 000 | 1000 | 500 | 250 | 125 |
| e | 8800 | 880 | 440 | 220 | 110 |

3

| ×5 | First multiply by 10 | Then halve it | Multiplication fact |
|---|---|---|---|
| a | 24 | 240 | 120 | 24 × 5 = 120 |
| b | 68 | 680 | 340 | 68 × 5 = 340 |
| c | 120 | 1200 | 600 | 120 × 5 = 600 |
| d | 500 | 5000 | 2500 | 500 × 5 = 2500 |
| e | 1240 | 12 400 | 6200 | 1240 × 5 = 6200 |

4

| ÷5 | First divide by 10 | Then double it | Division fact |
|---|---|---|---|
| a | 420 | 42 | 84 | 420 ÷ 5 = 84 |
| b | 350 | 35 | 70 | 350 ÷ 5 = 70 |
| c | 520 | 52 | 104 | 520 ÷ 5 = 104 |
| d | 900 | 90 | 180 | 900 ÷ 5 = 180 |
| e | 1200 | 120 | 240 | 1200 ÷ 5 = 240 |

5

| ×30 | First ×10 | Then ×3 | Multiplication fact |
|---|---|---|---|
| a | 15 | 150 | 450 | 15 × 30 = 450 |
| b | 22 | 220 | 660 | 22 × 30 = 660 |
| c | 33 | 330 | 990 | 33 × 30 = 990 |
| d | 150 | 1500 | 4500 | 150 × 30 = 4500 |
| e | 230 | 2300 | 6900 | 230 × 30 = 6900 |

6

| | × 30 | First × 3 | Then × 10 | Multiplication fact |
|---|---|---|---|---|
| a | 15 | 45 | 450 | 15 × 30 = 450 |
| b | 22 | 66 | 660 | 22 × 30 = 660 |
| c | 33 | 99 | 990 | 33 × 30 = 990 |
| d | 150 | 450 | 4500 | 150 × 30 = 4500 |
| e | 230 | 690 | 6900 | 230 × 30 = 6900 |

7 a 600 b 880 c 1250
  d 1700 e 840 f 5000
  g 1200 h 1440 i 570
  j $72 k $90 l 416
  m 208 n 62
8 $87.40

## Extended practice

1

| | × 15 | × 10 | Halve it to find × 5 | Add the two answers | Multiplication fact |
|---|---|---|---|---|---|
| a | 12 | 120 | 60 | 180 | 12 × 15 = 180 |
| b | 32 | 320 | 160 | 480 | 32 × 15 = 480 |
| c | 41 | 410 | 205 | 615 | 41 × 15 = 615 |
| d | 86 | 860 | 430 | 1290 | 86 × 15 = 1290 |
| e | 422 | 4220 | 2110 | 6330 | 422 × 15 = 6330 |

2

| | × 13 | × 10 | × 3 | Add the two answers | Multiplication fact |
|---|---|---|---|---|---|
| a | 15 | 150 | 45 | 195 | 15 × 13 = 195 |
| b | 12 | 120 | 36 | 156 | 12 × 13 = 156 |
| c | 23 | 230 | 69 | 299 | 23 × 13 = 299 |
| d | 31 | 310 | 93 | 403 | 31 × 13 = 403 |
| e | 25 | 250 | 75 | 325 | 25 × 13 = 325 |

3 a 2500 b 6300 c 4
  d 30 e 8800 f $1.80
  g $68 h $0.90
4 a 60 b 3600

## Unit 1: Topic 8

### Guided practice

1 a 162 b 325
2 a 508 b 981 c 630
  d 916 e 8190 f 9415
  g 8512 h 7285 i 9042

### Independent practice

1 a 700 b 540 c 1080
  d 1840 e 2010 f 2040
  g 3040 h 9840 i 10 980
  j 55 280 k 186 060 l 208 720
2 $4380
3 1 966 080
4 a 552 b 805 c 980
5 a 888 b 1053 c 1092

### Extended practice

1 a 725 b 1134 c 741
  d 1419 e 1125 f 2368
  g 3198 h 11 178 i 18 612
2 a 11 725 days b 83 700 sneezes

## Unit 1: Topic 9

### Guided practice

1 a 69 b 442 c 94
  d 110 e 4321 f 1201
  g 934 h 4841 i 4322
  j 4322 k 12 343 l 54 322
2 a 215 b 582 c 358
  d 2686 e 659 f 348

### Independent practice

1 a 4 r2 or $4\frac{2}{3}$ b 9 r2 or $9\frac{2}{5}$
  c 9 r3 or $9\frac{3}{4}$ d 8 r1 or $8\frac{1}{8}$
  e 8 r5 or $8\frac{5}{9}$ f 8 r5 or $8\frac{5}{7}$
  g 9 r3 or $9\frac{3}{9}$ or $9\frac{1}{3}$
  h 9 r4 or $9\frac{4}{6}$ or $9\frac{2}{3}$
2 a 116 r3 or $116\frac{3}{4}$ b 90 r2 or $90\frac{2}{3}$
  c 32 r5 or $32\frac{5}{6}$ d 148 r2 or $148\frac{2}{5}$
  e 858 r1 or $858\frac{1}{3}$ f 187 r1 or $187\frac{1}{9}$
  g 694 r1 or $694\frac{1}{6}$ h 331 r2 or $331\frac{2}{7}$
3 a 62 and one left over (Possible justification: they would not be able to split a marble in half.)
  b $7\frac{1}{2}$ (Possible justification: they would probably share the extra donut between them.)
4 a $26.50 b $18.50 c $11.50
  d $18.25 e $16.50 f $21.50
5 a 148.75 b 125.6 c 63.25
  d 136.8 e 336.75 f 1231.5
  g 2865.5 h 2319.6 i 6523.25
6 a 36 b $36.25

### Extended practice

1 a 291.33 b 124.25 c 83.14
  d 42.33 e 80.44 f 1828.33
  g 348.6 h 1226.14 i 11 494.75
  j 14 321.33 k 5095.2 l 10 807.22
2 a 2 b 2 c 4 d 3 e 2 f 5 g 5
3 It erupts 20 times a day. Look for the strategy that the student uses, e.g. division by estimation or 365 × 10 = 3650, double 3650 = 7300.

## Unit 1: Topic 10

### Guided practice

1

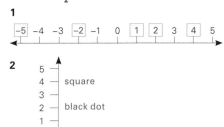

2

5 — 
4 — square
3 —
2 — black dot
1 —
0 —
-1 — triangle
-2 —
-3 — blue dot
-4 —
-5 — star

3 −5, −3, −1, 0, 2, 4

4 a true b false c true
  d false e true f false
  g false h true

### Independent practice

1 a −2 + 4 = 2 b 2 − 3 = −1
  c 4 − 7 = −3 d −6 + 5 = −1
  e −3 − 5 = −8 f −8 + 8 = 0
  g −8 + 10 = 2 h 7 − 11 = −4
  i −7 + 15 = 8 j 6 − 13 = −7
2 a −1 b −1 c −4
  d −5 e −10 f −60
3 a −60, −50, −40, −30, −20, −10, 0, 10, 20, 30, 40, 50
  b −25, −20, −15, −10, −5, 0, 5, 10, 15, 20, 25, 30
  c −28, −24, −20, −16, −12, −8, −4, 0, 4, 8, 12, 16
  d −35, −28, −21, −14, −7, 0, 7, 14, 21, 28, 35, 42
  e −63, −54, −45, −36, −27, −18, −9, 0, 9, 18, 27, 36
4 a & b Tuesday 2°C
      Sunday 1°C
      Wednesday 0°C
      Saturday −1°C
      Monday −2°C
      Friday −3°C
      Thursday −4°C
  c Thursday d 6°C

5

−5 −4 −3 −2 −1 0 1 2 3 4 5
 M   A              T         H S

### Extended practice

1 a Helsinki, Montreal, Quebec, Moscow
  b Berlin
  c −5°C
  d Montreal and Sydney (−6° and 27°), Helsinki and Acapulco (−3° and 30°), Quebec and Melbourne (−7° and 26°)

2

| | INTERNATIONAL BIG BANK | | |
|---|---|---|---|
| Date | Paid in $ | Paid out $ | Balance $ |
| 3 May | 100 | | 100 |
| 4 May | | 120 | −20 |
| 9 May | 30 | | 10 |
| 14 May | | 50 | −40 |
| 31 May | 45 | | 5 |

3 a $−100
  b Teacher to check, e.g. it would be more than $90 because there would be interest payable on the amount owing.

## Unit 1: Topic 11

### Guided practice

1

| | Multiplication | Base number and exponent |
|---|---|---|
| a | 2 × 2 × 2 × 2 × 2 | $2^5$ |
| b | 4 × 4 × 4 | $4^3$ |
| c | 8 × 8 × 8 × 8 | $8^4$ |
| d | 5 × 5 × 5 × 5 × 5 | $5^5$ |
| e | 7 × 7 × 7 × 7 × 7 × 7 | $7^6$ |
| f | 10 × 10 × 10 × 10 | $10^4$ |

2.

| | Base number and exponent | Number of times the base number is used in a multiplication | Multiplication | Value of the number |
|---|---|---|---|---|
| a | $3^3$ | three times | $3 \times 3 \times 3$ | 27 |
| b | $2^4$ | four times | $2 \times 2 \times 2 \times 2$ | 32 |
| c | $5^3$ | three times | $5 \times 5 \times 5$ | 125 |
| d | $6^2$ | two times | $6 \times 6$ | 36 |
| e | $9^2$ | two times | $9 \times 9$ | 81 |
| f | $10^3$ | three times | $10 \times 10 \times 10$ | 1000 |

3.

| | Starting number | What number multiplied by itself makes the number? | Square root of the starting number | Number fact |
|---|---|---|---|---|
| a | 4 | $2 \times 2 = 4$ | 4 | $\sqrt{4} = 2$ |
| b | 36 | $6 \times 6 = 36$ | 6 | $\sqrt{36} = 6$ |
| c | 9 | $3 \times 3 = 9$ | 3 | $\sqrt{9} = 3$ |
| d | 64 | $8 \times 8 = 64$ | 8 | $\sqrt{64} = 8$ |

4.

| | Starting number | Which two square numbers is it between? | What are their square roots? | The square number is between |
|---|---|---|---|---|
| a | 10 | 9 and 16 | $\sqrt{9} = 3$ and $\sqrt{16} = 4$ | 3 and 4 |
| b | 42 | 36 and 49 | $\sqrt{36} = 6$ and $\sqrt{49} = 7$ | 6 and 7 |
| c | 20 | 16 and 25 | $\sqrt{16} = 4$ and $\sqrt{25} = 5$ | 4 and 5 |
| d | 52 | 49 and 64 | $\sqrt{49} = 7$ and $\sqrt{64} = 8$ | 7 and 8 |

## Independent practice

1. a $2^7 = 2 \times 2 \times 2 \times 2 \times 2 \times 2 \times 2 =$
   b $5^5 = 5 \times 5 \times 5 \times 5 \times 5 = 3125$
   c $3^6 = 3 \times 3 \times 3 \times 3 \times 3 \times 3 = 729$
   d $4^5 = 4 \times 4 \times 4 \times 4 \times 4 = 1024$
   e $7^4 = 7 \times 7 \times 7 \times 7 = 2401$
2. a $8^5$: $9^4 = 6561$ and $8^5 = 32\,768$
   b $3^5$: $5^3 = 125$ and $3^5 = 243$
3. a $5^6 = 15\,625$
   b $10^6 = 1\,000\,000$

4.

| Starting number | The approximate square root is between | Actual square root (to two decimal places) | Number fact |
|---|---|---|---|
| 5 | 2 and 3 | 2.24 | $\sqrt{5} = 2.24$ |
| 40 | 6 and 7 | 6.32 | $\sqrt{40} = 6.32$ |
| 14 | 3 and 4 | 3.74 | $\sqrt{14} = 3.74$ |
| 30 | 5 and 6 | 5.48 | $\sqrt{30} = 5.48$ |
| 99 | 9 and 10 | 9.95 | $\sqrt{99} = 9.95$ |

## Extended practice

1. a $5^2 = 25$
   b $3^4 = 81$
   c $10^4 = 10\,000$
   d $1^{10} = 1$
2. Teachers may wish to discuss with students an appropriate number of decimal places that are needed.
   a $8^{-1} = 1 \div 8 = 0.125$
   b $8^{-2} = 1 \div 8 \div 8 = 0.015625$
   c $4^{-1} = 1 \div 4 = 0.25$
   d $4^{-2} = 1 \div 4 = 0.25 \div 4 = 0.0625$
   e $10^{-2} = 1 \div 10 = 0.1 \div 10 = 0.01$
   f $10^{-3} = 1 \div 10 = 0.1 \div 10 = 0.01 \div 10 = 0.001$
3. This is a useful discussion point. For square numbers, the method of saying "multiplied by itself" works. However, for higher exponents, such as the power of 4, it is not correct to say, "$3^4 = 3$ times itself 4 times" because this could be confused with $3 \times 3 \times 3 \times 3$.
   $3 \times 3$ (first time) = 9, $\times 3$ (second time) = 27, $\times 3$ (third time) = 81, $\times 3$ (fourth time) = 243. $3^4 = 81$. If dealt with correctly, $3^4$ is seen as the base number (4) being used in a multiplication 4 times: $3 \times 3 \times 3 \times 3 = 81$.
   Advanced student should, therefore, see $6^{-3}$ as the opposite of $6^{-3}$ in which 1 is divided by 6 three times ($1 \div 6 \div 6 \div 6$). Therefore, $6^3$ can be written as 1 multiplied by 6 three times ($1 \times 6 \times 6 \times 6$).
   a $6^3 = 1 \times 6 \times 6 \times 6 = 216$ (compare with $6 \times 6 \times 6 = 216$)
   b $4^4 = 1 \times 4 \times 4 \times 4 \times 4 = 256$ (compare with $4 \times 4 \times 4 \times 4 = 256$)
4. Some trial and error should lead students to see that the base number always remains the same when the exponent is 1. The method in question 3 is a simple way to see this. For example, $5^1 = 1 \times 5 = 5$ and so on.

# Unit 2: Topic 1

## Guided practice

1. a one-sixth, $\frac{1}{6}$     b three-fifths, $\frac{3}{5}$
   c seven-tenths, $\frac{7}{10}$   d two-thirds, $\frac{2}{3}$
2. a $\frac{4}{16}$ or $\frac{1}{4}$   b Student shades 4 stars.
3. a $\frac{9}{10}$
   b Student draws a smiley face at the $\frac{3}{10}$ mark.
   c $\frac{7}{10}$
   d Student draws a triangle at the $\frac{6}{10}$ mark.

## Independent practice

1. a $\frac{3}{6}, \frac{4}{8}, \frac{5}{10}, \frac{6}{12}$
   b Any fraction that is equivalent to $\frac{1}{2}$, e.g. $\frac{7}{14}, \frac{8}{16}$ etc.
2. The answers below are ones that students can identify from the fraction wall. However, there are other possibilities, such as $\frac{2}{10} = \frac{4}{20}$.
   a $\frac{1}{5}$     b $\frac{2}{12}$     c $\frac{1}{4}$
   d $\frac{10}{12}$   e $\frac{8}{10}$
   f Any or all of $\frac{1}{3}, \frac{2}{6}$ and $\frac{4}{12}$
3. a $\frac{2}{3}$   b $\frac{8}{10}$   c $\frac{6}{8}$
   d $\frac{6}{9}$   e $\frac{8}{12}$   f $\frac{3}{4}$
4. There are several possible correct responses. One possibility is shown. Check that the equivalent fractions are correct and that the shading matches the fractions.
   a $\frac{6}{8} = \frac{3}{4}$

   b $\frac{8}{10} = \frac{4}{5}$

5. Teacher to check shading and to decide how accurate students need to be with regard to dividing the shapes.
6. a $\frac{3}{12}$ (or any equivalent to $\frac{1}{4}$)
   b $\frac{3}{4}$ (or any equivalent to $\frac{3}{4}$)
   c Most likely answers are $\frac{1}{6}$ and $\frac{2}{12}$, but students could choose other equivalent fractions.
   d Student draws a star at the $\frac{8}{12}$ mark.

## Extended practice

1. a $\div 4$   b $\div 3$   c $\div 2$
   d $\div 3$   e $\div 5$   f $\div 4$
   g $\times 2$   h $\times 2$   i $\times 4$
   j $\times 2$   k $\times 6$   l $\times 4$
2. Teacher to check method. Below are probable solutions, but students may provide different answers, such as $\frac{4}{10}$ for 2d:
   a $\frac{2}{3}$   b $\frac{3}{4}$   c $\frac{1}{2}$
   d $\frac{2}{5}$   e $\frac{1}{7}$   f $\frac{4}{5}$
   g $\frac{1}{4}$
3. a $\frac{1}{2}$   b $\frac{4}{5}$   c $\frac{1}{3}$
   d $\frac{1}{3}$   e $\frac{1}{6}$   f $\frac{4}{5}$

## Unit 2: Topic 2

### Guided practice

1. a $\frac{4}{6}$ (or $\frac{2}{3}$)  b $\frac{8}{7} = 1\frac{1}{7}$
   c $\frac{6}{4} = \frac{3}{4}\frac{2}{4}$ (or $1\frac{1}{2}$)  d $\frac{8}{10}$ (or $\frac{4}{5}$)
2. a $\frac{3}{8} + = \frac{5}{8}$  b $\frac{3}{4} - \frac{2}{4} = \frac{1}{4}$

### Independent practice

1. a $\frac{5}{8}$   b $\frac{8}{10}$ (or $\frac{4}{5}$)   c $\frac{4}{7}\frac{5}{9}$
   d $\frac{5}{7}$   e $\frac{10}{12}$ (or $\frac{5}{6}$)   f $\frac{7}{9}$
   g $\frac{5}{6}$   h 1 (or $\frac{10}{10}$)   i $\frac{7}{8}$
2. a $\frac{4}{8}$ (or $\frac{1}{2}$)   b $\frac{6}{9}$ or $\frac{2}{3}$   c $\frac{6}{12}$ or $\frac{1}{2}$
   d $\frac{2}{4}$   e $\frac{2}{7}$   f $\frac{6}{10}$ or $\frac{2}{5}$
   g $\frac{4}{9}$   h $\frac{3}{6}$ or $\frac{1}{2}$
3. a $\frac{4}{8}$ or $\frac{1}{2}$   b $\frac{7}{10}$   c $\frac{3}{6}$ or $\frac{1}{2}$
   d $\frac{7}{8}$   e $\frac{6}{9}$   f $\frac{5}{9}$
4. a $\frac{1}{5}$   b $\frac{3}{10}$   c $\frac{4}{12}$ or $\frac{1}{3}$
   d $\frac{5}{9}$   e $\frac{2}{4}$   f $\frac{6}{12}$ or $\frac{1}{2}$
5. Student shows that $\frac{4}{9} + \frac{1}{3} = \frac{4}{9} + \frac{3}{9} = \frac{7}{9}$.
6. a $\frac{12}{8} = 1\frac{4}{8}$ or $1\frac{1}{2}$   b $\frac{12}{9} = 1\frac{3}{9}$ or $1\frac{1}{3}$
   c $\frac{16}{12} = 1\frac{4}{12}$ or $1\frac{1}{3}$   d $\frac{6}{4} = 1\frac{2}{4}$ or $1\frac{1}{2}$
   e $\frac{16}{10} = 1\frac{6}{10}$ or $1\frac{3}{5}$   f $\frac{9}{6} = 1\frac{3}{6}$ or $1\frac{1}{2}$
   g $\frac{8}{8} = 1$   h $\frac{4}{3} = 1\frac{1}{3}$
7. a $1\frac{4}{8}$ or $1\frac{1}{2}$   b $1\frac{1}{9}$
   c $1\frac{2}{10}$ or $1\frac{1}{5}$   d $2\frac{2}{4}$ or $2\frac{1}{2}$
   e $2\frac{6}{8}$ or $2\frac{3}{4}$   f $3\frac{9}{10}$
   g $1\frac{2}{9}$   h $2\frac{6}{8}$ or $2\frac{3}{4}$
8. a $\frac{9}{8}$ or $1\frac{1}{8}$   b $\frac{11}{10}$ or $1\frac{1}{10}$
   c $2\frac{3}{6}$ or $2\frac{1}{2}$   d $3\frac{3}{8}$
   e $2\frac{5}{10}$ or $2\frac{1}{2}$   f $4\frac{4}{9}$
9. a $\frac{7}{8}$   b $1\frac{8}{10}$ or $1\frac{4}{5}$   c $\frac{11}{12}$
   d $\frac{5}{6}$   e $1\frac{1}{4}$   f $\frac{5}{12}$

### Extended practice

1. Students should convert all fractions to eighteenths: $\frac{5}{18} + \frac{4}{18} + \frac{3}{18} + \frac{6}{18} = \frac{18}{18} = 1$
2. a $1\frac{1}{2}$   b $2\frac{5}{12}$   c $1\frac{5}{8}$
   d $1\frac{3}{10}$   e $5\frac{3}{4}$   f $1\frac{1}{2}$
   g $3\frac{7}{12}$
3. Answers will vary. For example, $\frac{1}{4} + \frac{1}{3} + \frac{5}{12} + \frac{1}{6} = \frac{14}{12} = 1\frac{1}{6}$.

## Unit 2: Topic 3

### Guided practice

1. a $\frac{3}{100}$ or 0.03   b $\frac{69}{100}$ or 0.69
   c $\frac{20}{100}$ ($\frac{2}{10}$) or 0.20 (0.2) (Discuss the connection with students.)
2. a 2   b 8   c 125
   d 200   e 75   f 9
   g 99   h 999   i 1
   j 10   k 100   l 250

### Independent practice

1. Student shades:
   a 5 squares   b 35 squares
   c 33 squares   d 90 squares
2. a True   b False   c True
   d False   e True   f True
   g False   h True   i True
   j False   k False   l True
3. Student shades:
   a 15 squares red   b 5 squares yellow

c 45 squares blue   d 10 squares green
e Unshaded amount is $\frac{1}{4}$ or 0.25.

4. 0.045, 0.145, 0.415, 0.45, 0.451

5. 
| | | |
|---|---|---|
| a | $\frac{3}{4}$ | 0.75 |
| b | $\frac{1}{10}$ | 0.1 |
| c | $\frac{3}{10}$ | 0.30 |
| d | $\frac{9}{100}$ | 0.09 |
| e | $\frac{405}{1000}$ | 0.405 |
| f | $\frac{250}{1000}$ | 0.25(0) |
| g | $\frac{99}{1000}$ | 0.099 |
| h | $\frac{1}{100}$ | 0.01 |

6. 
| | Improper fraction | Mixed number | Decimal |
|---|---|---|---|
| a | $\frac{7}{4}$ | $1\frac{3}{4}$ | 1.75 |
| b | $\frac{13}{10}$ | $1\frac{3}{10}$ | 1.3 |
| c | $\frac{125}{100}$ | $1\frac{25}{100}$ | 1.25 |
| d | $\frac{450}{100}$ | $4\frac{50}{100}$ (or equivalent) | 4.5 |
| e | $\frac{275}{100}$ | $2\frac{75}{100}$ (or equivalent) | 2.75 |
| f | $\frac{1250}{1000}$ | $1\frac{250}{1000}$ (or equivalent) | 1.25 |

### Extended practice

1. a 0.1   b 0.25   c 0.7
   d $\frac{1}{100}$   e $\frac{3}{4}$   f $\frac{1}{1000}$
2. a 0.2   b 0.125   c 0.75
   d 0.375   e 0.8   f 0.875
3. 0.33 recurring   4 0.1̇6̇
5. Teacher to check rounding. To 3 decimal places, 0.1428 becomes 0.143

## Unit 2: Topic 4

### Guided practice

1. a 4166   b 41.66
2. a 45.2   b 4.37   c 29.12
   d 52.3   e 1.75   f 26.97

### Independent practice

1. a 6.02   b 9.36   c 63.936
   d 50.1   e 1.55   f 7.593
   g 2.21   h 9.9   i 17.415
2. a $171.14   b $80.05
3. a 54.91 m   b 2.287 kg
4. d 8.253 seconds
5. 33.92 m
6. 10 kg
7. 8.61 m

### Extended practice

1. Answers will vary, e.g. 0.2 + 4.62 + 4.36 = 9.18 or 0.9 + 4.92 + 3.36 = 9.18

2. a 
| Organ | Mass |
|---|---|
| skin | 10.886 kg |
| liver | 1.56 kg |
| brain | 1.408 kg |
| lungs | 1.09 kg |
| heart | 0.315 kg |
| kidneys | 0.29 kg |
| spleen | 0.17 kg |
| pancreas | 0.098 kg |

b 1.405 kg   c 9.478 kg   d heart
e Teacher to check but the right must be 0.07 kg heavier and the two masses must be close (e.g. right: 0.58 kg, left: 0.51 kg)
f 0.992 kg   g 0.943 kg

## Unit 2: Topic 5

### Guided practice

1. a 396   b 39.6
2. a 8540   b 85.4(0)
3. a 2982   b 298.2
4. a 18 438   b 184.38
5. a 172   b 17.2
6. a 171   b 17.1
7. a 82   b 8.2
8. a 204   b 20.4

### Independent practice

1. a 44.1   b 85.6   c 63(.0)
   d 91.5   e 8.64   f 10.36
   g 8.1   h 7.71   i 51.8
   j 864.2   k 7.725   l 95.13
2. a 5.3   b 12.1   c 12.4
   d 19.5   e 1.05   f 2.37
   g 3.19   h 12.1   i 14.38
   j 12.47   k 2.57   l 0.527
3. a 174.4   b 327.6   c 36.95
   d 128.01   e 324.24   f 29.824
   g 722 (.0)   h 558.45   i 40.175
4. a $2   b $4   c $7.95   d $19.90
5. a $1   b $1.60   c $3   d $1.25

### Extended practice

Look for the strategies the student uses to solve these problems. Discuss whether mental strategies would be more appropriate in some cases.

1. $84.55
2. 60 cm (0.6 m)
3. $62.50
4. a 
| Item | Cost |
|---|---|
| Soft drink (1.25 L) | $6.75 |
| Juice (300 mL) | $5.04 ($5.05) |
| Potato crisps (50 g) | $16.20 |
| Chocolate (150 g) | $9.86 ($9.85) |
| Melon | $3.84 ($3.85) |
| Pies (4 in a pack) | (Two packs of four pies are needed for 6 students.) $16.08 ($16.10) |

b $57.80 if items are rounded or $57.77 without rounding. This would then be rounded to $57.75.
c 64 cents (round to 65 cents)
d $57.80 ÷ 6 = $9.65 (rounded)
   $57.77 ÷ 6 = $9.65 (rounded)
e $57.80 × 4 = $231.20
   $57.77 × 4 = $231.08 (= $231.10 rounded)

## Unit 2: Topic 6

### Guided practice

1. a 450  b 45
2. a 740  b 74
3. a 3750  b 375
4. a 6290  b 629
5. a 35  b 3.5
6. a 74  b 7.4
7. a 87  b 8.7
8. a 93  b 9.3
9. a 326  b 23.5
   c 78.92  d 652
10. a 2.35  b 4.275
    c 0.35  d 0.02

### Independent practice

1. a 35  b 350
2. a 67  b 670
3. a 53.8  b 538
4. a 40.9  b 409
5. a 0.45  b 0.045
6. a 0.79  b 0.079
7. a 5.45  b 0.545
8. a 6.27  b 0.627
9. a 245  b 1737
10. a 34.161  b 0.001
11. a 1300  b 2600  c 3570
    d 1270  e 15 470  f 72 950
    g 96 300  h 25 400
12. a 0.432  b 0.529  c 0.841
    d 0.697  e 1.485  f 3.028
    g 10.436  h 99.999

13.

|   | × 10 | × 100 | × 1000 |
|---|---|---|---|
| a | 1.7 | 17 | 170 | 1700 |
| b | 22.95 | 229.5 | 2295 | 22 950 |
| c | 3.02 | 30.2 | 302 | 3020 |
| d | 4.42 | 44.2 | 442 | 4420 |
| e | 5.793 | 57.93 | 579.3 | 5793 |
| f | 21.578 | 215.78 | 2157.8 | 21 578 |
| g | 33.008 | 330.08 | 3300.8 | 33 008 |
| h | 29.005 | 290.05 | 2900.5 | 29 005 |

14.

|   | ÷ 10 | ÷ 100 | ÷ 1000 |
|---|---|---|---|
| a | 74 | 7.4 | 0.74 | 0.074 |
| b | 7 | 0.7 | 0.07 | 0.007 |
| c | 18 | 1.8 | 0.18 | 0.018 |
| d | 325 | 32.5 | 3.25 | 0.325 |
| e | 2967 | 296.7 | 29.67 | 2.967 |
| f | 3682 | 368.2 | 36.82 | 3.682 |
| g | 14 562 | 1456.2 | 145.62 | 14.562 |
| h | 75 208 | 7520.8 | 752.08 | 75.208 |

### Extended practice

1. $225 \times 4 \div 1000$
2. a 0.936 ($312 \times 3 = 936$. Divide 936 by 1000 = 0.936)
   b 9.36 ($312 \times 3 = 936$. Divide 936 by 100 = 9.36)
   c 0.609 ($203 \times 3 = 609$. Divide 609 by 1000 = 0.609)
   d 8.004 ($4002 \times 2 = 8004$. Divide 8004 by 1000 = 8.004)
3. 2500 (Five jumps would get to one. 50 jumps would get to 10, 500 jumps would get to 100 so 2500 jumps would get to 500.)
4. a 600  b 750  c 1000
   d 250 ($150 \div 0.6$)
5. a $132.948 (rounded = $132.95)
   b $13 294.80
   c $1329.48 (rounded price = $1329.50)

## Unit 2: Topic 7

### Guided practice

1. a $\frac{9}{100}$, 0.9, 9%
   b $\frac{99}{100}$, 0.99, 99%
   c $\frac{80}{100}$ ($\frac{8}{10}$ or $\frac{4}{5}$), 0.8, 80%
   d $\frac{25}{100}$ ($\frac{1}{4}$), 0.25, 25%
   e $\frac{50}{100}$ ($\frac{1}{2}$), 0.5, 50%
   f $\frac{75}{100}$ ($\frac{3}{4}$), 0.75, 75%
2. a 0.02, 2%, student shades 2 squares
   b $\frac{20}{100}$ ($\frac{1}{5}$), 20%, student shades 20 squares
   c $\frac{35}{100}$ ($\frac{7}{20}$), 0.35, student shades 35 squares
   d 0.7, 70%, student shades 70 squares

### Independent practice

1.

|   | Fraction | Decimal | Percentage |
|---|---|---|---|
| a | $\frac{15}{100}$ (or equivalent) | 0.15 | 15% |
| b | $\frac{22}{100}$ (or equivalent) | 0.22 | 22% |
| c | $\frac{6}{10}$ (or equivalent) | 0.6 | 60% |
| d | $\frac{9}{100}$ | 0.09 | 9% |
| e | $\frac{9}{10}$ | 0.9 | 90% |
| f | $\frac{53}{100}$ | 0.53 | 53% |
| g | $\frac{1}{2}$ (or equivalent) | 0.5 | 50% |
| h | $\frac{1}{4}$ | 0.25 | 25% |
| i | $\frac{4}{100}$ (or equivalent) | 0.04 | 4% |
| j | $\frac{3}{4}$ | 0.75 | 75% |
| k | $\frac{1}{5}$ | 0.2 | 20% |

2. a True  b True  c False
   d True  e False  f False
   g False  h True  i True
3. a 20%, $\frac{1}{4}$, 0.3  b 0.07, $\frac{6}{10}$, 69%
   c $\frac{2}{100}$, 17%, 0.2  d 4%, 0.14, $\frac{1}{4}$
   e 10%, $\frac{1}{5}$, 0.5  f $\frac{3}{10}$, 39%, 0.395
4. Matching sets are:
   - 5%, 0.05, $\frac{1}{20}$
   - 8%, 0.08 and $\frac{8}{100}$
   - 80%, 0.8 and $\frac{8}{10}$
5. Percentages: 10%, 30%, 60%, 75%, 95%
   Fractions: $\frac{1}{10}$, $\frac{1}{4}$, $\frac{3}{4}$, $\frac{9}{10}$ (or equivalents)
6. a 5%  b 22%  c 44%
   d 59%  e 72%  f 99%
7. Student draws smiley face and arrow pointing to the point approximately mid-way between 0.8 and 0.9.
8. a Triangle: 30%  b Star: 40%
   c Circle: 70%  d Hexagon: 90%

### Extended practice

1. 2% of the world's cattle is in Australia.
2. $\frac{4}{5}$ of Australian mammal species are found nowhere else in the world.
3. The population of Victoria was approximately $5\frac{1}{2}$ million people in 2009.
4. 15% of the sheep in the Top 10 sheep countries are in Australia.
5. The Great Sandy Desert covers one-twentieth of Australia.
6. About 13% of the world's threatened animal species are in Australia.

## Unit 3: Topic 1

### Guided practice

1. a 8:4  b 3:6  c 9:3
2. a 2:1  b 1:2  c 3:1
   d 1:4  e 2:1  f 5:1
   g 1:3

### Independent practice

1. a 9  b 15  c 24
2. a 6  b 10  c 20
3. a 1:3:5  b 1:3:2  c 2:3:1
4. Students should recognise that the ratio of 3:1:2 means that there should be 3 blue squares for each yellow square and every 2 green squares. The simplest solution is to follow this pattern. However, students may choose to colour any 12 squares blue, any 4 squares yellow and any 8 squares green.
5. a Look for students who use appropriate vocabulary, such as, "For every yellow bead there are 3 red beads and 4 blue beads." By this stage, students should be beyond simply counting the number of beads in each colour.
   b 1:3:4
6. a & b Students might see that they can substitute the yellow beads in question 5 for green and continue the pattern. In this case, they could describe the pattern in the following way: "For every green bead there are 3 red beads and 4 blue beads." The ratio would be 1:3:4. Students requiring a greater challenge can be encouraged to use a different ratio, such as 1:2:5, with 3 green beads, 6 red beads and 15 blue beads.

7.

| Flour | Milk | Eggs | Number of pancakes |
|---|---|---|---|
| 120 g | 250 mL | 1 | 8 |
| 240 g | 500 mL | 2 | 16 |
| 480 g | 1000 mL or 1L | 4 | 32 |
| 720 g | 1.5 L | 6 | 48 |
| 60 g | 125 mL | $\frac{1}{2}$ | 3 |

8. a Students who have had experience if finding the highest common factor of a set of numbers will use this skill to work out the ratio of the animals as 3:8:1:2. Others will probably use a process of trial and error.
   b Having worked out the ratio, students should see that Zoe has 4 ducks, 6 sheep, 16 goats and 2 horses.

## Extended practice

1. **a** $\frac{3}{4}$  **b** 75%  **c** 0.75
2. Students may need to revisit the procedure for finding a fraction of a quantity where the numerator is larger than 1.
   - **a** 2 oranges, 8 apples
   - **b** 5 oranges, 20 apples
   - **c** 10 oranges, 40 apples
   - **d** 7 oranges, 28 apples
3. **a** There are five "portions" and each portion is one-fifth. One-fifth of 45 pieces of fruit is 9 pieces of fruit (45 ÷ 5 = 9). So, 3 × 9 = 27 oranges and 2 × 9 =18 apples.
   **b** There are seven "portions" and each portion is one-seventh. One-seventh of 56 pieces of fruit is 8 pieces of fruit (56 ÷ 7 = 8). So, 3 × 8 = 24 oranges and 4 × 8 = 32 apples.
   **c** There are four "portions" and each portion is one-quarter. One-quarter of 32 pieces of fruit is 8 pieces of fruit (32 ÷ 4 = 8). So, 1 × 8 = 8 oranges and 3 × 8 apples = 24 apples.
   **d** There are eight "portions" and each portion is one-eighth. One-eighth of 72 pieces of fruit is 9 pieces of fruit (72 ÷ 8 = 9). So, 3 × 9 = 27 oranges and 5 × 9 apples = 45 apples.

## Unit 4: Topic 1

### Guided practice

1. **a** 4 × 5 = 20  **b** 6 × 4 = 24
2. 

| Position | 1 | 2 | 3 | 4 | 5 | 6 | 7 | 8 | 9 |
|---|---|---|---|---|---|---|---|---|---|
| Number | 10 | 9.5 | 9 | 8.5 | 8 | 7.5 | 7 | 6.5 | 6 |

Rule: Subtract 0.5

3. **a** Yes (because 24 is divisible by 4).
   **b** Yes (because 16 is divisible by 4).
   **c** No (because 42 is not divisible by 4).
4. Teacher to check, e.g. *Is the last digit even?*

### Independent practice

1. Teacher to check, e.g. because only 3 sticks are needed for each square after the first one.
2. 

| | Number of sticks |
|---|---|
| a | 1 + 4 × 3 = 12 × 1 = 13 |
| b | 1 + 6 × 3 = 1 + 18 = 19 |
| c | 1 + 8 × 3 = 1 + 24 = 25 |

3. True
4. Teacher to check, e.g.
   **a** You use 3 sticks for every triangle.
   **b** You start with 1 stick and then use 2 sticks for every triangle or start with 3 sticks for the first triangle and then use 2 for every other triangle.
   **c** You use 6 sticks for every hexagon.
   **d** You start with 1 stick and then use 5 sticks for every hexagon or start with 6 sticks for the first hexagon and then use 5 for every other hexagon.

5. 

| Position | 1 | 2 | 3 | 4 | 5 | 6 | 7 | 8 | 9 | 10 |
|---|---|---|---|---|---|---|---|---|---|---|
| Number | 2 | 5 | 8 | 11 | 14 | 17 | 20 | 23 | 26 | 29 |

6. 

| Position | 1 | 2 | 3 | 4 | 5 | 6 | 7 | 8 | 9 | 10 |
|---|---|---|---|---|---|---|---|---|---|---|
| Number | 1 | 4 | 9 | 16 | 25 | 36 | 49 | 64 | 81 | 100 |

Rule: Teacher to check, e.g. to find the number, you square the position number or to find the number, you multiply the position number by itself.

7. Depending on the experience and ability levels of the students, teachers may wish to discuss and work through the flow chart and model an example to show how the chart works. The division examples allow for various pathways through the flow chart.
   **a** 114 r2  **b** 137  **c** 86 r1
8. Teachers may wish to use this activity as a cooperative group activity. The flow chart is a basic one. A successfully completed chart may look like this:

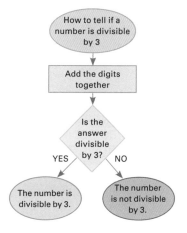

### Independent practice

1. **a** 5 × 2 = 2 + 8   **b** 5 × 5 = 30 – 5   **c** 24 ÷ 2 = 4 × 3
   **d** 20 + 7 = (4 + 5) × 3   **e** $\frac{1}{2}$ of 6 + 5 = 24 ÷ 3

2. 

| Problem | | Split the problem to make it simpler | | Solve the problem | | Answer |
|---|---|---|---|---|---|---|
| e.g. 27 × 3 | = | (20 × 3) + (7 × 3) | = | 60 + 21 | = | 81 |
| a  23 × 4 | = | (20 × 4) + (3 × 4) | = | 80 + 12 | = | 92 |
| b  19 × 7 | = | (10 × 7) + (9 × 7) | = | 70 + 63 | = | 133 |
| c  48 × 5 | = | (40 × 5) + (8 × 5) | = | 200 + 40 | = | 240 |
| d  37 × 6 | = | (30 × 6) + (7 × 6) | = | 180 + 42 | = | 222 |
| e  29 × 5 | = | (20 × 5) + (9 × 5) | = | 100 + 45 | = | 145 |
| f  43 × 7 | = | (40 × 7) + (3 × 7) | = | 280 + 21 | = | 301 |
| g  54 × 9 | = | (50 × 9) + (4 × 9) | = | 450 + 36 | = | 486 |

3. 

| Problem | | Change the order to make it simpler | | Solve the problem | | Answer |
|---|---|---|---|---|---|---|
| e.g. 20 × 17 × 5 | = | 20 × 5 × 17 | = | 100 × 17 | = | 1700 |
| a  20 × 13 × 5 | = | 20 × 5 × 13 | = | 100 × 13 | = | 1300 |
| b  25 × 14 × 4 | = | 25 × 4 × 14 | = | 100 × 14 | = | 1400 |
| c  5 × 19 × 2 | = | 5 × 2 × 19 | = | 10 × 19 | = | 190 |
| d  25 × 7 × 4 | = | 25 × 4 × 7 | = | 100 × 7 | = | 700 |
| e  60 × 12 × 5 | = | 60 × 5 × 12 | = | 300 × 12 | = | 3600 |
| f  5 × 18 × 2 | = | 5 × 2 × 18 | = | 10 × 18 | = | 180 |
| g  25 × 7 × 8 | = | 25 × 8 × 7 | = | 200 × 7 | = | 1400 |

## Extended practice

1. Teachers may wish to discuss the use of a formula in order to simplify a rule before students start this work.
   **a** 32  **b** 40  **c** 80  **d** 200
2. **a** 10
   **b** $n = 2 + t \times 2$ or $n = t \times 2 + 2$ (because, apart from the two end tables, only two sides of each rectangle can be used). Note, the above formulae follow the pattern of the other formulae in this topic. However, other formulae are possible, such as $n = 6 + (t - 2) \times 2$. Teacher to decide whether discussion on this is appropriate.
   **c** a) 18; b) 22; c) 42; d) 102
3. **a** 7  **b** 9  **c** 12  **d** 22
4. **a** Teacher to check formula, e.g. $n = t + 2$ or $n = 2 + t$ (because, apart from the two end tables, only one side of each triangle can be used). See also note in question 2b.
   **b** 22

## Unit 4: Topic 2

### Guided practice

1. 17
2. **a** 7  **b** 10  **c** 21
   **d** 6  **e** 4  **f** 8
   **g** 14  **h** 10
3. **a** 12  **b** 12  **c** 6
   **d** $4\frac{1}{2}$  **e** 21  **f** 29
   **g** 12  **h** 36
4. **a** 15  **b** 10  **c** 8
   **d** 2  **e** 100  **f** 18
   **g** 20  **h** 30

**4**

| | Problem | Use opposites | Find the value of ◊ | Check by writing the equation |
|---|---|---|---|---|
| e.g. | ◊ + 15 = 35 | ◊ = 35 − 15 | ◊ = 20 | 20 = 35 − 15 |
| a | ◊ × 6 = 54 | ◊ = 54 ÷ 6 | ◊ = 9 | 9 × 6 = 54 |
| b | ◊ + 1.5 = 6 | ◊ = 6 − 1.5 | 4.5 | 4.5 + 1.5 = 6 |
| c | $\frac{1}{4}$ of ◊ = 10 | ◊ = 10 × 4 | 40 | $\frac{1}{4}$ of 40 = 10 |
| d | ◊ × 10 = 45 | ◊ = 45 ÷ 10 | 4.5 | 4.5 × 10 = 45 |
| e | ◊ ÷ 10 = 3.5 | ◊ = 3.5 × 10 | 35 | 35 ÷ 10 = 3.5 |
| f | ◊ ÷ 4 = 1.5 | ◊ = 1.5 × 4 | 6 | 6 ÷ 4 = 1.5 |
| g | ◊ × 100 = 725 | ◊ = 725 ÷ 100 | 7.25 | 7.25 × 100 = 725 |

**5**

| | Problem | Possible substitutes for ◊ | | | | Check |
|---|---|---|---|---|---|---|
| e.g. | ◊$^2$ × 3 = 75 | 4 | 5 | 6 | 7 | 5$^2$ × 3 = 25 × 3 = 75 |
| a | ◊ × 3 + 5 = 32 | 8 | 9 | 10 | 11 | 9 × 3 + 5 = 32 |
| b | 54 ÷ ◊ − 5 = 1 | 9 | 10 | 11 | 12 | 54 ÷ 9 − 5 = 1 |
| c | 2 × ◊ + 5 = 15 | 2 | 3 | 4 | 5 | 2 × 5 + 5 = 15 |
| d | 15 ÷ ◊ − 1.5 = 0 | 5 | 10 | 15 | 20 | 15 ÷ 10 − 1.5 = 0 |
| e | 24 × 10 − ◊ = 228 | 12 | 14 | 16 | 18 | 24 × 10 − 12 = 228 |
| f | ◊ ÷ 2 = 4$^2$ + 3 | 35 | 36 | 37 | 38 | 38 ÷ 2 = 16 + 3 |
| g | (5 + ◊) × 10 = 25 × 3 | 1.5 | 2 | 2.5 | 3 | (5 + 2.5) × 10 = 75 |

### Extended practice

**1 a** ◊ × 3 − 4 = 11, ◊ × 3 = 11 + 4, ◊ = 15 ÷ 3, ◊ = 5
 **b** ◊ × 10 − 15 = 19, ◊ × 10 = 19 + 15, ◊ × 10 = 34, ◊ = 34 ÷ 10, ◊ = 3.4

**2 a** 16
 **b** Answers may vary, e.g. (10 + 2) × 4 − 2 = 46 or 10 + 2 × 4 − 2 = 16
 **c** Answer will depend on the calculator used. Most basic calculators will not be programmed to follow the order of operations and will give the answer 46.
 **d** Teacher to check, e.g. (10 + 2) × 4 − 2 = 46, 10 + 2 × (4 − 2) = 14, (10 + 2) × (4 − 2) = 24

**3** Answers will vary, e.g. 4 − 4 + 4 − 4 = 0; (4 ÷ 4 − 4) + 4 = 1; 4 ÷ 4 + 4 ÷ 4 = 2; (4 + 4 + 4) ÷ 4 = 3; (4 − 4) × 4 + 4 = 4; (4 × 4 + 4) ÷ 4 = 5; (4 + 4) ÷ 4 + 4 = 6; 4 + 4 − 4 ÷ 4 = 7; 4 − 4 + 4 + 4 = 8; 4 ÷ 4 + 4 + 4 = 9

## Unit 5: Topic 1

### Guided practice

**1**

| 4 km | 4000 m |
|---|---|
| 7 km | 7000 m |
| 19 km | 19 000 m |
| 6 km | 6000 m |
| 7.5 km | 7500 m |
| 3.5 km | 3500 m |
| 4.25 km | 4250 m |
| 9.75 km | 9750 m |

**2**

| 1 m | 100 cm |
|---|---|
| 4 m | 400 cm |
| 5.5 m | 550 cm |
| 2.5 m | 250 cm |
| 7.1 m | 710 cm |
| 8.2 m | 820 cm |
| 1.56 m | 156 cm |
| 0.75 m | 75 cm |

**3**

| 5 cm | 50 mm |
|---|---|
| 42 cm | 420 mm |
| 9 cm | 90 mm |
| 3.2 cm | 32 mm |
| 7.5 cm | 75 mm |
| 12.5 cm | 125 mm |
| 12.4 cm | 124 mm |
| 9.9 cm | 99 mm |

**4** Teacher to check and possibly ask student to justify answers.
 **a** cm or mm **b** cm
 **c** mm **d** m
 **e** m or cm **f** km

### Independent practice

**1** Answers may vary. Teachers could ask students to justify answers.

| a | The length of a pencil | 157 mm | 15.7 cm |
|---|---|---|---|
| b | The height of a Year 6 student | 1.57 m | 157 cm |
| c | The length of a finger nail | 15 mm | 1.5 cm |
| d | The distance around a school yard | 157 m | 0.157 km |
| e | The length of a bike ride | 1570 m | 1.57 km |

**2**

| | mm | cm and mm | cm |
|---|---|---|---|
| a | 45 mm | 4 cm 5 mm | 4.5 cm |
| b | 75 mm | 7 cm 5 mm | 7.5 cm |
| c | 82 mm | 8 cm 2 mm | 8.2 cm |
| d | 69 mm | 6 cm 9 mm | 6.9 cm |

**3** Teacher to check. Note: accuracy in measuring is less important than the ability to convert between the units of length.

**4 a** Teacher to check estimates. Look for estimates expressed in the correct unit and that are reasonable in comparison with the length of Line B.
 **b** Line A: 6.8 cm, Line C: 9.2 cm

**5 a** Teacher to check estimates. Look for estimates that are reasonable given the length of Line B.
 **b** All lines are 6 cm long.

**c** Teacher may choose to discuss optical illusions. This particular optical illusion is called the Müller-Lyer illusion and there are various theories as to why we perceive the lines as different in length.

**6** Allow for a tolerance of +/− 4 mm per rectangle and 3 mm for the triangle. Allow equivalent lengths.
 **a** 10 cm **b** 11.6 cm
 **c** 9.2 cm **d** 6.3 cm

**7** Teacher to check.

### Extended practice

**1**

| Tyrannosaurus Rex | 12.8 m | 2 |
|---|---|---|
| Iguanodon | 6800 mm | 3 |
| Microraptor | 0.83 m | 6 |
| Homalocephale | 290 cm | 5 |
| Saltopus | 590 mm | 7 |
| Puertasaurus | 3700 cm | 1 |
| Dromiceiomimus | 3500 mm | 4 |
| Micropachycephalosaurus | 50 cm | 8 |

**2** Answers will vary. Look for students who come up with a plausible suggestion such as a dog or a cat.

**3** Puertasaurus

**4** Answers will vary, e.g. about 26 (37 ÷ 1.4 = 26.43). Look for students who can make a reasonable estimate of the height of Year 6 students and use this to come up with a plausible response.

**5** Answers will vary. Look for students who are able to make a reasonable estimate of their own height and to accurately calculate the difference between their given height and that of the microraptor.

**6** Teacher to check and decide on level of accuracy that is required. Looking at problem-solving strategies may be seen to be more important than absolute accuracy.

## Unit 5: Topic 2

### Guided practice

**1** 8 cm$^2$
**2** 12 cm$^2$
**3** 10 cm$^2$
**4 a** 2 **b** 2 **c** 4 cm$^2$
**5 a** 2 cm$^2$ **b** 9 cm$^2$ **c** 18 cm$^2$

### Independent practice

**1 a** L: 3 cm, W: 2 cm, A: 6 cm$^2$
 **b** L: 3 cm, W: 5 cm, A: 15 cm$^2$
**2 a** 40 m$^2$ **b** 63 m$^2$ **c** 150 m$^2$
**3 a** 21 m$^2$ **b** 56 m$^2$ **c** 24 m$^2$
**4** Students' own answer. Look for students who show an understanding of why the formula would not work, e.g. because the shape is not a rectangle.
**5 a** 20 cm$^2$ **b** 25 cm$^2$ **c** 18 cm$^2$
 **d** 16 cm$^2$ **e** 16 cm$^2$
**6** 5000 m$^2$
**7 a** 20 000 m$^2$ **b** 40 000 m$^2$
 **c** 50 000 m$^2$
**8** 30 cm × 21 cm = 630 cm$^2$

## Extended practice

1. **a** ABCD = 20 cm², ABC = 10 cm²
   **b** EFGH = 18 cm², EFG = 9 cm²
   **c** IJKL = 21 cm², JKL = 10.5 (10½) cm²
   **d** MNOP = 16 cm², NOQ = 8 cm²
2. **a** 12 cm²
   **b** 7.5 (7½) cm²
   **c** 12.5 (12½) cm²

# Unit 5: Topic 3

## Guided practice

1. **a** 8 cm³  **b** 16 cm³
   **c** 12 cm³  **d** 12 cm³
2. **a** 12 cm³, 3, 36 cm³
   **b** 8 cm³, 2, 16 cm³
3. **a**

   | 3 kL | 3000 L |
   |---|---|
   | 9 kL | 9000 L |
   | 3.5 kL | 3500 L |
   | 6.25 kL | 6250 L |

   **b**

   | 2 L | 2000 mL |
   |---|---|
   | 7 L | 7000 mL |
   | 5.75 L | 5750 mL |
   | 4.5 L | 4500 mL |

   **c**

   | 500 cm³ | 500 mL |
   |---|---|
   | 225 cm³ | 225 mL |
   | 1000 cm³ | 1 L |
   | 1750 cm³ | 1750 mL or 1.75 L |

## Independent practice

1. **a** 15   **b** 15 cm³
2. Teacher to check. Look for students who are able to find the correct answer using a reliable strategy, e.g. because 6 cm³ would fit on the top layer and there are two layers the same.
3. Teacher to check, e.g. you multiply the width by the length to find the number of cubes that will fit on the one layer and then multiply that by the height to find the volume. (V = L × W × H)
4. **a** 40 cm³  **b** 18 cm³  **c** 48 cm³
   **d** 48 cm³  **e** 80 cm³  **f** 180 cm³
5. B, C, A, D, F, G, E
6. **a** 1400 mL  **b** 1500 mL  **c** 1300 mL
   **d** 1250 mL  **e** 750 mL  **f** 1350 mL
   Teacher to check that the shading is appropriate.

## Extended practice

1. The most likely answer is millimetres. Students could be asked to justify their responses.
2. 30 m × 3 m × 0.15 m = 13.5 m³.
3. Teacher to check. Look for students who are able to accurately follow the directions and who make the link between volume and capacity to arrive at a plausible answer. Note: equipment in primary schools is not usually accurate enough to prove that 1 mL water has a volume of exactly 1 cm³. Teachers may choose to discuss this with students.

# Unit 5: Topic 4

## Guided practice

1. **a**

   | 5 t | 5000 kg |
   |---|---|
   | 7.5 t | 7500 kg |
   | 1.25 t | 1250 kg |
   | 2.355 t | 2355 kg |
   | 0.995 t | 995 kg |

   **b**

   | 3.5 kg | 3500 g |
   |---|---|
   | 4.5 kg | 4500 g |
   | 0.85 kg | 850 g |
   | 0.25 kg | 250 g |
   | 3.1 kg | 3100 g |

   **c**

   | 5.5 g | 5500 mg |
   |---|---|
   | 3.75 g | 3750 mg |
   | 1.1 g | 1100 mg |
   | 0.355 g | 355 mg |
   | 0.001 g | 1 mg |

2. **a–d** Multiple possible answers, e.g. a: a truck, b: a person, c: flour, d: a grain of salt. Look for students who make appropriate choices for each unit of mass and who can justify their answers.

3. 

   | | kg and fraction | kg and decimal | kg and g |
   |---|---|---|---|
   | a | 3½ kg | 3.5 kg | 3 kg 500 g |
   | b | 2½ kg | 2.5 kg | 2 kg 500 g |
   | c | 3¼ kg | 3.25 kg | 3 kg 250 g |
   | d | 4 7/10 kg | 4.7 kg | 4 kg 700g |
   | e | 1 9/10 kg | 1.9 kg | 1 kg 900 g |

## Independent practice

1. **a** 1 kg 700 g (or equivalent)
   **b** 4 kg 250 g (or equivalent)
   **c** 850 g (or equivalent) (Allow a tolerance of +/– 10 g)
2. Most likely answers are below. Look for students who can explain why they would choose the particular scale to find the mass of each item.
   **a** Scale C
   **b** Scale B
   **c** Scale A
   **d** Scale B or C
3. The scale has 50 g increments so the pointer should be between 900 and 950. Teacher to decide on the required level of accuracy.

4. **a** B, A, C, D
   **b** A & B (exactly 5 t)
   **c** C & D (5.945 t)

5. Teacher to check. Look for students who are able to describe the relationship between the total mass of the pad and the mass of each sheet, e.g. find the mass of 100 sheets and divide the answer by 100.
6. Check that the total equals 1.85 kg and also that the masses are appropriate. (For example, one item at 1.844 kg and the remaining six items at 1 g would not be appropriate.)
7. **a** 62.5 kg
   **b** Twelve Year 6 students. (12 × 40 kg = 480 kg, 13 × 40 kg = 520 kg)
8. Check that total equals 1 kg and also that the masses are appropriate. (For example, one mango at 985 g and the remaining three at 5 g would not be appropriate.)

## Extended practice

1. Teacher to check. Look for students who are able to come up with a strategy that connects millilitres and grams and demonstrates their understanding of mass. It is unlikely for normal primary classroom equipment to be accurate enough to prove that 1 mL of water has a mass of 1 g. This could be a useful discussion point for students.
2. **a** potato crisps
   **b** breakfast cereal has 40 mg more sodium. (However, students could be asked to consider the normal serving size.)
   **c** 505 mg (2 × 135 mg + 55 mg + 180 mg)
   **d** 216 mg (Total = 2516 mg less 2300 mg = 216 mg)

# Unit 5: Topic 5

## Guided practice

1. **a** 1:20 pm, 1320
   **b** 6:48 pm, 1848
   **c** Clock to show 2:42, 0242
   **d** 11:07 pm, 2307
   **e** Clock to show 10:22, 10:22 pm
   **f** 6:27 am, 0627
   **g** Clock to show 10:35, 1035
   **h** Clock to show 11:59, 11:59 pm

## Independent practice

1. 54 minutes
2. Train 8221
3. 9 minutes
4. Train 8215
5. Because the train only stops there to pick up passengers.
6. 55 minutes
7. 

   | Station | Time |
   |---|---|
   | Southern Cross | 1630 |
   | Footscray | 1638 |
   | Werribee | 1656 |
   | Little River | 1704 |
   | Lara | 1710 |
   | Corio | 1714 |
   | North Shore | 1716 |
   | North Geelong | 1720 |
   | Geelong | 1724 |

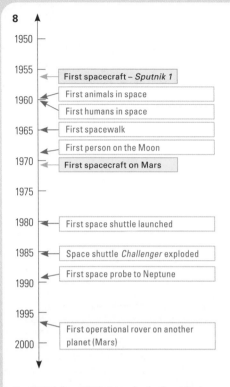

**9** 1971 (allow 1972 at teacher's discretion)

### Extended practice

1. **a** 3 hours **b** 2 hours 51 minutes
   **c** 9 minutes
2. **a** 3:26 pm **b** Clock to show 3:26
   **c** 22 minutes
3. 

| | Departs Big Town | Arrives Small Town |
|---|---|---|
| Bus A | 1208 | 1507 |
| Bus B | 1533 | 1832 |
| Bus C | 1954 | 2253 |

## Unit 6: Topic 1

### Guided practice

1. **a** regular hexagon
   **b** irregular quadrilateral
   **c** regular quadrilateral (square)
   **d** irregular pentagon
   **e** regular octagon
   **f** regular pentagon
   **g** irregular triangle
   **h** irregular hexagon
2. It has five equal sides and some of the angles are the same size.

### Independent practice

1. Teacher to check descriptions. For example:
   **a** Equilateral triangle. All the angles are the same size. All the sides are the same length.
   **b** Isosceles triangle. Two angles are the same size. Two sides are the same length.
   **c** Right-angled scalene triangle. All the sides are different lengths. One angle is a right angle.
   **d** Right-angled isosceles triangle. Two sides are the same length. One angle is a right angle.
   **e** Scalene triangle. All sides are different lengths. All angles are different sizes.
2. Teacher to check descriptions. For example:
   **a** Square. All the angles are right angles. All the sides are the same length.
   **b** Trapezium. There are two obtuse angles. There is one pair of parallel sides.
   **c** Rectangle. All the angles are right angles. Two pairs of sides are the same length.
   **d** Parallelogram. There are two pairs of parallel sides. There are two obtuse and two acute angles.
   **e** Rhombus. All the sides are the same length. There are two pairs of parallel sides.
3. Teacher to check. Examples of similarities and differences that can be observed:

| | Similarities | Differences |
|---|---|---|
| a | Neither shape has any straight lines. | The diameters on the circle are all the same length but they are different on the oval. |
| b | They are both regular shapes. | One has five sides and the other has eight sides. |
| c | They each have at least one pair of parallel sides | The parallelogram has two pairs of parallel sides but the trapezium has only one. |
| d | They are both irregular hexagons. | One has three pairs of parallel sides but the other has only one pair. |
| e | They are both pentagons. | One is regular but the other is not. |
| f | They are both parallelograms. | One (the rhombus) has all sides of equal length but the other does not. |
| g | They are both quadrilaterals. | One has obtuse and acute angles and the other has four right angles. |
| h | They are both right-angled triangles. | The first is a scalene triangle but the second is isosceles. |
| i | They both have a right angle. | The first shape has a reflex angle. |
| j | They are both octagons. | The first is a regular octagon but the second octagon is not. |
| k | They both have at least four reflex angles. | The first shape has 10 sides but the second shape has 8 sides. |

### Extended practice

1. **a** circumference **b** radius
   **c** diameter
2. **a** sector **b** quadrant
   **c** semi-circle
3. Teacher to check and to decide on the level of accuracy that is appropriate.
4. 8

## Unit 6: Topic 2

### Guided practice

1. **a** rectangular prism
   **b** square (-based) pyramid
   **c** triangular prism
   **d** cylinder
   **e** octagonal prism
   **f** hexagonal pyramid
   **g** square prism
   **h** cone
   **i** triangular pyramid

### Independent practice

1. **a** rectangular prism
   **b** square (-based) pyramid
2. Teacher to check. Note: it may be necessary to discuss the reason for adding tags to some of the faces. The page could be enlarged by photocopying. Alternatively, students could copy the nets onto grids of a larger size.
3. Teacher to check. Note: teachers will probably wish to provide extra paper for additional practice and to reassure students that success is not necessarily assured at the first attempts in such activities. Look for students who show an understanding of the faces and edges of 3D shapes and are able to accurately reproduce them on isometric dot paper.

### Extended practice

1. 

| | Name | Number of faces | Number of vertices | Number of edges | Does Euler's Law work? |
|---|---|---|---|---|---|
| a | rectangular prism | 6 | 8 | 12 | Yes (14 − 12 = 2) |
| b | hexagonal prism | 8 | 12 | 18 | Yes (20 − 18 = 2) |
| c | square (-based) pyramid | 5 | 5 | 8 | Yes (10 − 8 = 2) |
| d | triangular prism | 5 | 6 | 9 | Yes (11 − 9 = 2) |
| e | triangular pyramid | 4 | 4 | 6 | Yes (8 − 6 = 2) |
| f | square prism | 6 | 8 | 12 | Yes (14 − 12 = 2) |
| g | pentagonal prism | 7 | 10 | 15 | Yes (17 − 15 = 2) |
| h | octagonal prism | 10 | 16 | 24 | Yes (26 − 24 = 2) |

## Unit 7: Topic 1

### Guided practice

1. **a** 80°, acute **b** 100°, obtuse
   **c** 35°, acute **d** 145°, obtuse
2. Teacher to check that angles of 25° are drawn. (Decide whether to allow a tolerance of $x°$.) Look for students who understand how to align the centre of the protractor correctly along the baseline of the angle and who understand which set of numbers on the protractor to read.

### Independent practice

1. (Allow a tolerance of +/− 1°)
   **a** 125° **b** 165° **c** 99° **d** 169°
2. Teacher to check, e.g. The size of the "outside angle" is 40° less than 360°.
3. **a** 330° **b** 315° **c** 215° **d** 265°
4. **a** 95° **b** 112° **c** 270° **d** 120°
   **e** 333° **f** 40° **g** 30° **h** 120°
   **i** 45° **j** 155°

## Extended practice

1. **a** $a = 60°$, $b = 180°$
   **b** $a = 125°$, $b = 55°$, $c = 125°$
   **c** $a = 48°$, $b = 132°$, $c = 132°$
   **d** $a = 50°$
2. $b = 38°$, $c = 142°$, $d = 38°$
   $e = 38°$, $f = 142°$, $g = 38°$, $h = 142°$
   $i = 142°$, $j = 38°$, $k = 142°$, $l = 38°$,
   $m = 38°$, $n = 142°$, $o = 38°$, $p = 142°$
3. Teacher to check and decide on level of accuracy that is required. Looking at problem-solving strategies may be seen to be more important than absolute accuracy.

# Unit 8: Topic 1

## Guided practice

1. **a** reflection **b** translation **c** rotation
2. Teacher to check patterns. Look for students who are able to correctly identify and continue the transformation used in each pattern.

## Independent practice

1. Teacher to check descriptions. For example:
   **a** The hexagon has been translated horizontally.
   **b** The triangle has been rotated horizontally.
   **c** The hexagon has been translated vertically.
   **d** The pentagon has been reflected vertically.
   **e** The triangle has been reflected horizontally and vertically.
   **f** The arrow has been translated horizontally. The second row is the same as the first one, but it has been reflected horizontally.
2. Teacher to check pattern. Look for students who can accurately use the language of transformation to describe the pattern.
3. Teacher to check patterns. (Shapes are coloured to simplify the identification of the patterns.) Examples of possible descriptions:
   **a** The shape has been reflected horizontally on the first row. The second row is a vertical reflection of the first row.
   **b** The shape is rotated through 180° clockwise on the first row (or has been reflected horizontally then vertically on the first row). The second row is a vertical translation of the first row.
   **c** The shape is translated horizontally on the first row. The second row is a vertical reflection of the first row.
4. Teacher to check pattern. Look for students who are able to successfully demonstrate their understanding of transformations through the construction of their pattern.

## Extended practice

**1 & 2** Teacher to check. Look for students who are able to demonstrate proficiency with digital technologies to construct a pattern that draws on their understanding of transformations.

# Unit 8: Topic 2

## Guided practice

1. yellow triangle
2. **a** (−8,−6) **b** (4,4)
3. green circle and yellow triangle
4. True
5. **a** The student draws a line from the origin point to the yellow triangle.
   **b** (1,−1), (2,−2), (3,−3)
   **c & d** Teacher to check. Look for students who demonstrate an understanding of the quadrant system and who interpret the coordinates correctly.

## Independent practice

1. **a** (3,5) **b** (−4,5) **c** (−4,1) **d** (3,1)
2. (−6,5) → (−7,3)
3. Starting point students' choice. Endpoint must be the same as the starting point, e.g. (−4,1) → (3,1) → (3,5) → (−4, 5) → (−4,1)
4. Teacher to check drawing and coordinates. (It can be an advantage to ask students to give each other their plotted points to check that the drawings match the ordered pairs.) Look for students who demonstrate an understanding of how ordered pairs work and who can use them to accurately describe the points of their figure.
5.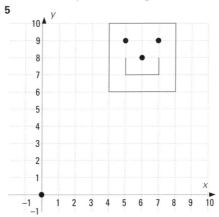
6. Teacher to check.

## Extended practice

1.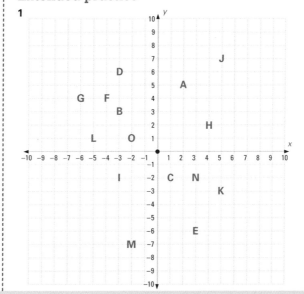

# Unit 9: Topic 1

## Guided practice

1. **a** 42 **b** 6
2. 2
3. Accept from $310 to $315.
4. 9

## Independent practice

1. **a** Teacher to check students' graphs.
   **b** 27 (Red & blue = 50. Yellow and purple = 23. 50 − 23 = 27)

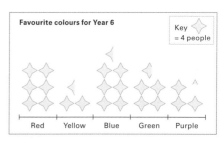

2. Teacher to check information. Totals must be 6 more than the information from question 1.
3. **a** Teacher to check graph. The most appropriate increment for the vertical axis is 3, as the highest possible bar total is 32.
   **b** Answers will vary, e.g. a bar graph because it is easier to work out the numbers for each bar, or a pictograph because it looks better.

**4 a**

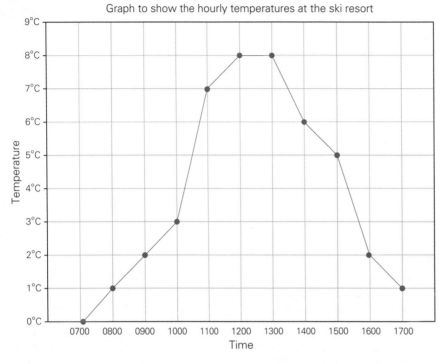

**b** Teacher to check. Look for students who comment on the information (rather than on the appearance of the graph). For example, the lowest temperature was at 7 am, OR the temperature remained the same between 1200 and 1300 (not "I like the way the lines go up and down").

**5** Teacher to check. Look for the way that the student labels the graph and, if necessary, whether an appropriate scale has been used. An appropriate way to present the information would be on a dot plot, a bar graph or a pictograph. Students who choose a line graph have not understood that this is used to show how recorded data changes over a period of time.

### Extended practice

**1 a** Annabelle, Jade and Eva
**b** Approximately one-sixteenth
**2 a** Numbers for each name
**b** Any two numbers around 600 that have a difference of 14. (The actual numbers were 612 named Eva and 598 named Annabelle.)
**3** Teacher to check, e.g. **Similarities**: Half of each graph is taken up with one name. The proportion of the second most popular name to the most popular name is about the same on each graph (half). There are two names on each graph that are about as popular as each other.
**Differences**: The fraction of the second most popular name is slightly smaller for girls than it was for boys.
**Note for teachers**: The information was taken from a Victorian government website. The data was deliberately chosen to make the graphs look similar. The names are all from the top 100, but the positions of the names in the top 100 are different on each. For example, the top boys' name (Jack) was deliberately not included, whereas the top girls' name was. See more at https://online.justice.vic.gov.au/bdm/popular-names.

**4 a** Any number between 1550 and 1649. (Actual number was 1596).
**b** Teacher to check, e.g. there were more than 1500 more babies with the top name for boys than there were for girls.

## Unit 9: Topic 2

### Guided practice

**1 a** secondary **b** primary
**2** Some questions may have two possible answers. Teachers may wish students to justify their responses. Likely answers are:
  **a** sample **b** census
  **c** sample or census (depending on size of school)
  **d** sample
**3 a** primary **b** sample

### Independent practice

**1** Secondary data (the class teacher did the survey).
**2 a** $5 (\frac{5}{8})$ **b** $1 (\frac{1}{8})$
**3** Yes, because the principal wrote that it was the majority of the students who were surveyed.
**4** It could be true (because it is not clear what the other students would have answered).
**5** … and could be true. Teachers may wish to open up a discussion about this as it raises issues about data in the media. Many may disagree with the newspaper headline but we cannot be sure what the majority of students in the town think (although neither could the newspaper editor!).

**6 a** About $130 to $135
  **b** four
  **c** About $42
**7 a** sample
  **b** Teacher to check e.g. "No, because it was only based on the views of 200 people". This issue could be used as a basis for a group or class discussion.
  **c** 180 ($\frac{9}{10}$ of 200)

### Extended practice

**1 a** sample **b** secondary
**2 a** Any number over 50
  **b** 49%
  **c** Teacher to check, e.g. the public has a right to know how many people were surveyed and what type of people they were.
**3 a** Teacher to check. Look for students who show an understanding of data collection and the importance of accurately representing data and its sources; e.g. the people surveyed (all students) were not a fair representation of public opinion.
  **b** Teacher to check. Look for students who show an understanding of how data collection influences results and can apply this to the situation, e.g. carry out a survey of 100 people of various ages and backgrounds.
  **c** Teacher to check, e.g. it is based on truth, but it cannot necessarily be trusted as a fair reflection of public opinion. This is a possible group or class discussion point.
**4 a** Teacher to check. Look for students who can offer reasoning to demonstrate their understanding of manipulation of survey responses, e.g. because it probably influenced the answers that people gave.
  **b** Teacher to check. Look for students who choose a question that is likely to result in the collection of more accurate data, e.g. Do you think a fast-food restaurant should be opened near the high school?

## Unit 9: Topic 3

### Guided practice

**1 a** Range is 16% to 80% = 64%
  **b** Range is 75 cm to 150 = 75 cm
**2 a** 35% **b** 76
**3 a** 44% **b** 16
**4 a** 30% **b** 15

# Independent practice

**1**

| Week | Seven-day minimum temperatures | Order | Range | Mode | Median | Mean |
|---|---|---|---|---|---|---|
| 1 | 3°, 6°, 7°, 9°, 7°, 8°, 2° | 2°, 3°, 6°, 7°, 7°, 8°, 9° | 7° | 7° | 7° | 6° |
| 2 | 1°, 3°, 2°, 9°, 7°, 7°, 6° | 1°, 2°, 3°, 6°, 7°, 7°, 9° | 7° | 7° | 6° | 5° |
| 3 | 9°, 6°, 8°, 8°, 10°, 7°, 8° | 6°, 7°, 8°, 8°, 8°, 9°, 10° | 4° | 8° | 8° | 9° |
| 4 | 10°, 9°, 10°, 8°, 7°, 3°, 2° | 2°, 3°, 7°, 8°, 9°, 10°, 10° | 8° | 10° | 8° | 9° |

**2**

| | Number set | Order | Range | Median |
|---|---|---|---|---|
| a | 8, 2, 6, 4, 10 | 2, 4, 6, 8, 10 | 8 | 6 |
| b | 25, 14, 17, 12, 6, 4 | 4, 6, 12, 14, 17, 25 | 21 | 13 |
| c | 12, 8, 2, 6, 2, 5, 21 | 2, 2, 5, 6, 8, 12, 21 | 19 | 6 |
| d | 82, 23, 3, 8, 15, 3, 16, 2 | 2, 3, 3, 8, 15, 16, 23, 82 | 80 | 11.5 or 11½ |

**3**

| | Number set | Mode | Mean |
|---|---|---|---|
| a | 8, 2, 6, 4, 10 | None | 6° |
| b | 25, 14, 17, 12, 6, 4 | None | 13 |
| c | 12, 8, 2, 6, 2, 5, 21 | 2 | 8 |
| d | 82, 23, 3, 8, 15, 3, 16, 2 | 3 | 19 |

**4 a** This could be done as a 'think, pair, share' activity, with students sharing their thoughts before arriving at their responses.

The modes should be fairly easy to estimate as (8 hours for Sydney and 6 hours for London).

Looking at the high and low points for each city is likely to result in answers of around 6 or 7 hours of sunshine a day for Sydney and 3, 4 or 5 hours a day of sunshine for London.

**b**

| | Sydney | London |
|---|---|---|
| Mode | 8 hours | 6 hours |
| Mean | 88 ÷ 12 = 7.33 (rounded to 7 hours) | 50 ÷ 12 = 4.166 (rounded to 4 hours) |

**c** This could take the form of a class or group discussion. Answers will vary. Responses will likely revolve around the difficulty to accurately estimate the mean without doing a calculation.

**d** Sydney. The mode (8) is only slightly more than the median (7.5) compared to London, where the mode (6) is 1.5 hours different to the median (4.5).

**e** London. The mean for October to March is 16 ÷ 6 = 2.66 hours a day, rounded to 3 hours. The mean for April to September is 34 ÷ 6 = 5.66 hours a day, rounded to 6 hours. The difference is an average of 3 hours a day less sunshine in the colder months.

**f** Sydney. The mean for October to March is 46 ÷ 6 = 7.66 hours a day, rounded to 8 hours. The mean for April to September is 42 ÷ 6 = 7 hours a day. The difference is 1 hour a day less sunshine in the colder months.

# Extended practice

**1 a** Yes, technically Sam is correct because 10 occurs more frequently than the other scores.

**b** Answers will vary, teacher to check. A possible answer is that more than half of the scores are less than 10, with two of the scores being very low.

**c** The median score is 7 out of 10.

**d** The mean score is 6 out of 10. Teachers may choose to ask students to reflect on whether Sam's achievement level is best reflected by the mean, median or mode.

**2 a** The range is 19 (20 − 1 + 19).

**b** 19    **c** 19

**d** The mean score is 155 ÷ 10 = 15.5 or $15\frac{1}{2}$.

**e** Teachers may choose to use this task for a group discussion about why the mean score does not reflect Sam's ability. The score of 1 out of 20 could be for a variety of reasons ranging from lack of effort to not feeling very well. When interpreting data in the real world, an anomaly (or outlier) is often ignored in order to give a truer interpretation of the data. Students could be further extended by carrying out a similar activity for their own assessments.

**3** This could be carried out as a group activity. Multiple answers are possible. Look for students who total the four temperatures (108) and subtract this from 203 (7 × 29). The answer of 95 needs to be divided appropriately between the three remaining days. For example, 31 °C, 32 °C and 32 °C, instead of 93 °C, 1 °C and 1 °C.

# Unit 10: Topic 1

## Guided practice

**1** Teacher to check. Answers may vary and students could be asked to justify their answers. Probable answers are:
- **a** even chance
- **b** highly likely
- **c** impossible
- **d** likely
- **e** certain
- **f** highly unlikely
- **g** unlikely

**2** $\frac{1}{10}$ (1 out of 10)

**3** 50%

**4** 0.3

## Independent practice

**1** 15%

**2** Students may choose a fraction, a decimal and a percentage in any order but possible answers are:
- **a** $\frac{2}{10}$ (or $\frac{1}{5}$)    **b** 0.4    **c** 10%

**3** There are 8 out of 10 ways the spinner will not land on green. Answers should be any or all of $\frac{8}{10}$, $\frac{4}{5}$, 0.8 or 80%.

**4** Teacher to check the appropriateness of student responses. Look for students who demonstrate an understanding of the language and application of probability and who are able to justify their responses.

**5** Teacher to check. Sectors should be coloured as follows:
- yellow: 2 sectors
- blue: 3 sectors
- green: 2 sectors
- white: 2 sectors
- red: 1 sector

**6 a** 0.8    **b** $\frac{7}{10}$    **c** 0.07
   **d** $\frac{4}{10}$    **e** $\frac{3}{4}$    **f** 8%

**7** $\frac{4}{10}$

**8** 2 should be red, 4 should be yellow and 6 should be blue

**9** A: 25 blue & 75 yellow
B: 60 blue & 40 yellow
C: 90 blue & 10 yellow
D: 50 blue & 50 yellow

## Extended practice

**1 a** 37    **b** $37
**c** Answers may vary, e.g. because the boss only gives back $36 of the $37.
**d** $1000

**2 a** Answers may vary, e.g. 18 out of 37 is almost the same as 18 out of 36, and 18 out of 36 = $\frac{1}{2}$.
**b** 19
**c** Students' own responses, e.g. because $37 was collected but the boss only paid back $36. Look for students who understand the probability of landing on black and can apply this to supply a plausible response.
**d** $10 000

# Unit 10: Topic 2

## Guided practice

1. **a** $\frac{5}{6}$ or 5 out of 6
   **b** Students' own responses. Look for students who demonstrate an understanding of probability and the fact that, although there is a greater chance of not rolling a 6, it is still possible, e.g. each number has the same chance so there is as much chance for 6 as for every other number.

2. Answers will vary. This could prove an interesting group or class discussion point, with students being asked to justify their responses. Look for students who can explain why different students obtained different results using the language of probability.

3. The probability for each number is 6. The likelihood of this occurring is probably not very high given the relatively small number of rolls of the dice. This could prove an interesting group or class discussion point about what would be likely to happen after, say, 360 or 3600 rolls of the dice.

## Independent practice

1. **a** 2
   **b** one (1 + 1)

2. 

| Total of two dice | Ways the dice can land | Total number of ways |
|---|---|---|
| 12 | 6 + 6 | 1 |
| 11 | 6 + 5, 5 + 6 | 2 |
| 10 | 6 + 4, 4 + 6, 5 + 5 | 3 |
| 9 | 6 + 3, 3 + 6, 5 + 4, 4 + 5 | 4 |
| 8 | 6 + 2, 2 + 6, 5 + 3, 3 + 5, 4 + 4 | 5 |
| 7 | 6 + 1, 1 + 6, 5 + 2, 2 + 5, 4 + 3, 3 + 4 | 6 |
| 6 | 5 + 1, 1 + 5, 4 + 2, 2 + 4, 3 + 3 | 5 |
| 5 | 4 + 1, 1 + 4, 3 + 2, 2 + 3 | 4 |
| 4 | 3 + 1, 1 + 3, 2 + 2 | 3 |
| 3 | 2 + 1, 1 + 2 | 2 |
| 2 | 1 + 1 | 1 |

3. 7 (6 out of 36 ways)

4. Allow fractional equivalents of the following:
   **a** $\frac{2}{36}$   **b** $\frac{3}{36}$   **c** $\frac{4}{36}$
   **d** $\frac{5}{36}$   **e** $\frac{6}{36}$   **f** $\frac{5}{36}$
   **g** $\frac{4}{36}$   **h** $\frac{3}{36}$   **i** $\frac{2}{36}$
   **j** $\frac{1}{36}$   **k** $\frac{0}{36}$

5. Probable number of times for each total:
   12: 2    11: 4    10: 6    9: 8
   8: 10    7: 12    6: 10    5: 8
   4: 6     3: 4     2: 2

6. Answers may vary, but likely answers are:
   Spinner 1: $\frac{3}{20}$ yellow, $\frac{11}{20}$ blue, $\frac{6}{20}$ red;
   Spinner 2: $\frac{1}{12}$ yellow, $\frac{7}{12}$ blue, $\frac{4}{12}$ red.

## Extended practice

1. **a** 1 out of 7    **b** one counter

2–3. Teachers may wish to model this game with students and discuss the implications of the game, which demonstrates why the only sure, long-term winner in a gambling situation is the "banker". It may be necessary to allow more than ten rounds of the game to establish a pattern. Teachers will also decide whether to "tweak" the rules so that each player must choose a different number each time. In this case the banker's balance is certain to increase by one counter each round. However, if, for example, each of the seven students "bets" on the same number and the spinner lands on that number, the "banker" will obviously lose. In the long term, however, the probabilities of the game will ensure that the only certain winner is the "banker".